T0214535

# SpringerBriefs in Mathematics

**SpringerBriefs in Mathematics** showcases expositions in all areas of mathematics and applied mathematics. Manuscripts presenting new results or a single new result in a classical field, new field, or an emerging topic, applications, or bridges between new results and already published works, are encouraged. The series is intended for mathematicians and applied mathematicians.

More information about this series at http://www.springer.com/series/10030

Nadia Mazza

# Endotrivial Modules

 Springer

Nadia Mazza
Department of Mathematics and Statistics
Lancaster University
Lancaster, UK

ISSN 2191-8198          ISSN 2191-8201   (electronic)
SpringerBriefs in Mathematics
ISBN 978-3-030-18155-0       ISBN 978-3-030-18156-7   (eBook)
https://doi.org/10.1007/978-3-030-18156-7

Mathematics Subject Classification (2010): 20C05, 20C20, 19A22, 20D15, 20C30, 20C33, 20C34, 20D30

This Springer imprint is published by the registered company Springer Nature Switzerland AG
The registered company address is: Gewerbestrasse 11, 6330 Cham, Switzerland

# Preface

These notes aim to provide a self-contained comprehensive account of *Endotrivial Modules*, collecting in one place most of the results that have been published in numerous journals and book chapters since the late 1970s. The endotrivial modules are deemed to be "important" in the modular representation theory of finite groups, and we give some motivation in support of this claim, mainly in the first two chapters. They were singled out in the late seventies because they were identified as the "building blocks" of the sources of simple modules for a large family of finite groups. But prior to this, they were identified as the "invertible" elements in the stable module category of finitely generated modules for a finite group, thus providing information about the self-equivalences of this category. In fact, the concept of an endotrivial module makes sense in greater generality, e.g. in many finite-dimensional Hopf algebras, or in categories equipped with a suitable tensor product and identity element [41, 103, 117].

In Chap. 1, we provide the reader with a concise review of the theoretical background needed, and we set the specific context and our notation. In particular, we review the notions needed in finite group theory, modular representation theory, homological algebra, and, in view of recent developments in the study of endotrivial modules, we also need to include some concepts from algebraic topology.

Chapter 2 starts with the definition of endotrivial and endo-permutation modules. Given a finite group $G$ and a field $k$ of positive characteristic $p$ with $p$ dividing the order of $G$, a finitely generated $kG$-module $M$ is *endotrivial* if $\operatorname{End}_k M$ is isomorphic to the trivial $kG$-module in the stable module category; that is, $\operatorname{End}_k M \cong k \oplus (\text{proj})$ as $kG$-modules, where $k$ denotes the trivial $kG$-module and (proj) a projective $kG$-module. It turns out that the set of stable isomorphism classes of endotrivial $kG$-modules forms a group with respect to the binary operation given by the tensor product of the modules over the ground field $k$.

We present the subject following the historical development of the theory of endotrivial and endo-permutation modules, building on E. Dade's seminal article [53], in which he singled out some "useful" modules that could be made into abelian groups. Thereafter, in 1990 L. Puig proved that for finite $p$-groups, the groups of endo-permutation and endotrivial modules are *finitely generated* [107], and this was extended for endotrivial modules to all finite groups in 2006 [34]. These milestones would suggest that a classification of endotrivial modules for finite groups may be possible.

In Chap. 3, we present some general results on the classification of endotrivial modules, or more precisely, the description of the group $T(G)$ of endotrivial modules for a finite group $G$. In particular, we discuss the unavoidable "lifting" question from $k$ to $\mathscr{O}$, where $(F, \mathscr{O}, k)$ is a $p$-modular system. That is: "*Given an endotrivial kG-module M, is there an endotrivial $\mathscr{O}$G-module $\tilde{M}$ such that $M \cong k \otimes_{\mathscr{O}} M$ is the reduction mod p of $\tilde{M}$?*". This question has a positive answer, implying that it suffices to classify endotrivial $kG$-modules, as they always lift. Next, we describe the structure of $T(G)$ for all finite groups with a normal Sylow $p$-subgroup, with a cyclic Sylow $p$-subgroup, with a dihedral, semi-dihedral, and a generalised quaternion Sylow 2-subgroup, and we end with the description of $T(G)$ for a $p$-soluble group $G$. In Chap. 3 we use traditional modular representation theory in order to find $T(G)$, and we also discuss the key obstacles to a complete classification of endotrivial modules for any finite group $G$ and any characteristic. By "traditional" methods, we mean the standard tools of (tensor) induction and restriction, and the theory of vertices and sources of indecomposable modules, recalled in Sects. 1.3 to 1.5; in contrast to the methods inspired by ordinary character theory, algebraic geometry and algebraic topology, which we use in Sects. 5.1 to 5.3.

In Chap. 4, we discuss "the torsionfree part" of the group $T(G)$ of endotrivial modules of a finite group $G$ over a field of characteristic $p$. By "the torsionfree part", we mean "a torsionfree abelian subgroup $TF(G)$ of $T(G)$ such that $T(G) = TT(G) \oplus TF(G)$", where $TT(G)$ is the torsion subgroup of $T(G)$ (which is finite because $T(G)$ is finitely generated). The $\mathbb{Z}$-rank of $TF(G)$ only depends on the $p$-subgroup structure of $G$ and on the action of $G$ by conjugation on its noncyclic elementary abelian $p$-subgroups. So, we make a detour via the category of noncyclic elementary abelian $p$-subgroups of a finite group, and show some of its properties. In particular, for endotrivial modules, the results give an upper bound on the maximal $p$-rank of $G$, such that beyond this bound $TF(G)$ must be cyclic, and an upper bound on the maximal $\mathbb{Z}$-rank of $TF(G)$. We end the chapter with results about finding generators for $TF(G)$ when $TF(G)$ is not cyclic, and present various partial results, including very recent ones obtained by Barthel, Grodal and Hunt, who use algebraic topology to describe $TF(G)$.

Chapter 5 handles the most difficult open question about $T(G)$. Namely, the problem of finding all (indecomposable) trivial source endotrivial modules for a finite group. The Green correspondence is the obvious first tool, but it does not suffice in general and several ingenious methods have been manufactured to try and find an answer: using ordinary character theory, methods derived from algebraic geometry, and methods using homotopy theory. We present each of these and give a few examples of their successful applications. Still, as we shall see, none of these methods is a panacea as we must rely at some point on known properties of a given group to find its group of endotrivial modules.

In Chap. 6, we collate the known results to date about the classification of endotrivial modules for "Very Important Groups", that is, symmetric and alternating groups and their covering groups, finite groups of Lie type, and sporadic simple groups and their covering groups. This final chapter ends with an idiosyncratic observation, leading to an open question. There are several outstanding problems, such as giving a presentation of the group of endotrivial modules by generators and relations when the torsion free rank is greater than one, and also describing the torsion group of the group of endotrivial modules for some Very Important Groups. Presently, research is ongoing to address these questions.

For the benefit of the reader, we have included a lengthy list of references (not exhaustive) which should help to fill in any omitted theoretical background, and also an alphabetical index, whose purpose is to help the reader navigate through the text. The numbering of the results is the standard one used by the Editor, of the form "chapter number.number", with a different counter for each of definitions, theorems, propositions, and lemmas in particular.

Finally, the author would like to thank all those who have worked with her on the quest for endotrivial modules: Serge Bouc, Jon Carlson, George Glauberman, Jesper Grodal, Dave Hemmer, Caroline Lassueur, Markus Linckelmann, Dan Nakano, Jacques Thévenaz, and also her former Ph.D. student Alec Gullon, and all the colleagues and friends whom she has met in her serendipitous mathematical journey over the last 15 years.

Lancaster, UK                                                                          Nadia Mazza
March 2019

# Contents

# Chapter 1
# Introduction

## 1.1 Introduction and Background

Throughout, $G$ denotes a finite group and $p$ a prime dividing the order $|G|$ of $G$. We want to study a certain class of *modular representations* of $G$ in characteristic $p$, called *endotrivial modules*. These modules are deemed "important" as they appear in the $p$-local description of sources of simple modules for finite $p$-soluble groups, and they also provide useful information about the self-equivalences of the *stable module category* of any finite group.

This first chapter aims to concisely review the theoretical notions required to study endotrivial modules, and suggests to the reader some selected literature already available on the subject for the proofs of the results stated.

## 1.2 Group Theory

In this short section, we include some concepts in group theory that we will use throughout. Suggested references are [65, 72, 73, 81, 109]. 

Let $G$ be a finite group and $p$ a prime divisor of the order $|G|$ of $G$. We write $\mathrm{Syl}_p(G)$ for the set of Sylow $p$-subgroups of $G$. The notation $(H < G)$ $H \leq G$ says that $H$ is a (proper) subgroup of $G$, and $(H \lhd G)$ $H \trianglelefteq G$ indicate that $H$ is a (proper) normal subgroup of $G$. Given $H, K \leq G$, we write $H \leq_G K$ if $H$ is *$G$-subconjugate to $K$*, i.e. if there exists a $g \in G$ such that $gHg^{-1} \leq K$. We write $|G : H|$ for the index of $H$ in $G$, i.e. the number of (left) *cosets* $G/H = \{gH \mid g \in G\}$ of $H$ in $G$, and $[G/H]$ denotes a set of representatives of the (left) cosets of $H$ in $G$. Unless otherwise specified, cosets are left cosets throughout. If $H, K \leq G$, then $K \backslash G / H = \{KgH \mid g \in G\}$ and $[K \backslash G / H]$ denote the double cosets and a set of their representatives, respectively. We write $A - B$ for the set difference $\{a \in A \mid a \notin B\}$ of two sets $A$ and $B$.

© The Author(s), under exclusive license to Springer Nature Switzerland AG 2019
N. Mazza, *Endotrivial Modules*, SpringerBriefs in Mathematics,
https://doi.org/10.1007/978-3-030-18156-7_1

Given a subgroup $H$ of $G$, we let $N_G(H)$ and $C_G(H)$ denote the normaliser and centraliser of $H$ in $G$, respectively. In particular, $Z(G) = C_G(G)$ is the *centre* of $G$ and a subgroup of $G$ is called *central* if it is contained in $Z(G)$. For any subgroup $H$ of $G$ we have an injective group homomorphism $N_G(H)/C_G(H) \rightarrow \mathrm{Aut}(H)$, given by $gC_G(H) \mapsto \left(c_g : h \mapsto ghg^{-1} \ (h \in H)\right)$, conjugation by $g$. We write $\mathrm{Aut}_G(H) = N_G(H)/C_G(H)$, and $\mathrm{Out}_G(H) = N_G(H)/HC_G(H)$ for the subgroup and subquotient of the automorphism group of $H$ induced by conjugation by the elements in $N_G(H)$.

Given elements $g, h \in G$, we write ${}^g h = ghg^{-1}$ and $h^g = g^{-1}hg$ for the conjugation of $h$ by $g$, and similarly ${}^g H = gHg^{-1}$ and $H^g = g^{-1}Hg$ for the conjugate subgroup of $H$ by $g$. We let $[g, h] = ghg^{-1}h^{-1}$ denote the *commutator* of $g, h \in G$, and $[H, K] = \langle [h, k] \mid h \in H, k \in K \rangle$ the subgroup of $G$ generated by the commutators of the elements of $H$ and $K$. We call $[G, G]$ the *derived subgroup* of $G$ and write $G'$ or $\gamma_2(G)$ instead. This is the smallest normal subgroup of $G$ with abelian factor group. That is, if $K \trianglelefteq G$ and $G/K$ is abelian, then $K \geq G'$. We say that $G$ is *perfect* if $G = [G, G]$.

The *Frattini subgroup* of $G$ is the intersection $\Phi(G)$ of all the maximal subgroups of $G$.

The centre, derived subgroup and Frattini subgroup of $G$ are instances of *characteristic* subgroups, i.e. they are invariant under any automorphism of $G$.

A nontrivial group $G$ is *simple* if $G$ has no proper nontrivial normal subgroups. We say that $G$ is *quasi-simple* if $G$ is perfect and the factor group $G/Z(G)$ is simple. For example, most special linear groups $\mathrm{SL}_n(q)$ are quasi-simple. We say that $G$ is *almost simple* if $G$ has a unique proper normal subgroup $G_0$ with $G_0$ simple nonabelian. For example, all symmetric groups $\mathfrak{S}_n$ for $n \geq 5$ are almost simple.

A group $G$ is *soluble* if its *derived series*

$$G > G' = [G, G] \geq \cdots \geq G^{(j)} = [G^{j-1}, G^{j-1}] \geq \cdots \geq 1$$

converges to 1.

A group $G$ is *nilpotent* if its *lower central series*

$$G > G' = [G, G] \geq \cdots \geq \gamma_{i+1}(G) = [\gamma_i(G), G] \geq \ldots \gamma_c(G) > \gamma_{c+1}(G) = 1$$

converges to 1, or equivalently its *upper central series*

$$1 < Z(G) = Z_1(G) \leq \cdots \leq Z_i(G) \leq \cdots \leq Z_{c-1}(G) < G = Z_c(G)$$

converges to $G$, where $Z_i(G)$ is the *$i$-th centre*, defined inductively as the unique subgroup of $G$ containing $Z_{i-1}(G)$ and such that $Z_i(G)/Z_{i-1}(G) = Z(G/Z_{i-1}(G))$ starting with $Z_1(G) = Z(G)$. The index $c$ appearing above is the *nilpotency class* of $G$. The upper and lower central series of $G$ are *central series* of $G$, that is, each subgroup $X_i$ is normal in $G$ and the successive quotients $X_i/X_{i+1}$ are central subgroups of $G/X_{i+1}$. Nilpotent groups are soluble and are characterised by the fact that they are the direct product of their Sylow $p$-subgroups, that is,

$$G = \prod_{p \mid |G|} O_p(G) \quad \text{where } O_p(G) \text{ is the largest normal } p\text{-subgroup of } G.$$

We also record here that every finite $p$-group is nilpotent, and that $N_G(H) > H$ for any proper subgroup $H$ of a nilpotent group $G$.

A group $G$ is $p$-*soluble* if $G$ has a normal series in which each factor group is either a $p$-group or a $p'$-group. Equivalently, this means that the series

$$1 \le O_p(G) \le O_{p,p'}(G) \le O_{p,p',p}(G) \le \ldots \quad \text{reaches } G,$$

where $O_{p,p'}(G)$ is the normal subgroup of $G$ containing $O_p(G)$ and such that

$$O_{p,p'}(G)/O_p(G) = O_{p'}(G/O_p(G)).$$

One can show that $G$ is soluble if and only if $G$ is $p$-soluble for every prime number $p$. A $p$-soluble group $G$ is $p$-*nilpotent* if $G = O_{p'}(G) \rtimes S$ for $S \in \mathrm{Syl}_p(G)$. Recall that by the *Schur–Zassenhaus theorem*, any extension

$$1 \longrightarrow N \longrightarrow G \longrightarrow K \longrightarrow 1 \quad \text{with } (|N|, |K|) = 1 \text{ and } N \trianglelefteq G \text{ splits,}$$

i.e. $G$ is isomorphic to the semi-direct product $G \cong N \rtimes K$.

A subgroup $H$ of $G$ is *trivial intersection*, or TI for short, if $H \cap {}^g H = 1$ for every element $g \in G - H$. A subgroup $H$ of $G$ is *strongly $p$-embedded* if $|G : H|$ and $|H \cap {}^g H|$ are coprime to $p$ for every element $g \in G - H$. In particular, $S$ is a TI subgroup of $G$ if and only if $N_G(S)$ is strongly $p$-embedded in $G$.

The *central product* of two groups $G$, $H$ over $Z$, for an abelian group $Z$ isomorphic to central subgroups of $G$ and of $H$ (without loss of generality, $Z \le Z(H) \cap Z(G)$), is the group

$$G *_Z H = (G \times H)/\langle (gz, h) = (g, zh) \mid g \in G, \ h \in H, \ z \in Z \rangle.$$

We often omit $Z$ and write $G * H$ if there is no possible confusion.

The wreath product $H \wr W$ for groups $H$ and $W \le \mathfrak{S}_d$, where $\mathfrak{S}_d$ is the symmetric group on $d$ letters, is the semi-direct product

$$(H_1 \times \cdots \times H_d) \rtimes W, \quad \text{where } H_i \cong H \text{ for each } 1 \le i \le d,$$

of $d$ copies of $H$ on which $W$ acts by permutation of the $H_i$'s:

$$(h_1, \ldots, h_d; \sigma)(g_1, \ldots, g_d; \tau) = (h_1 g_{\sigma^{-1}(1)}, \ldots, h_d g_{\sigma^{-1}(d)}; \sigma\tau).$$

For instance, $C_p \wr C_p$ is isomorphic to a Sylow $p$-subgroup of $\mathfrak{S}_{p^2}$, and $C_2 \wr C_2 \cong D_8$ is dihedral of order 8.

By the Sylow theorems, every finite group has a Sylow $p$-subgroup for every prime dividing $|G|$, and the Sylow $p$-subgroups of a given finite group $G$ are all $G$-conjugate. Amongst the multitude of theorems in group theory, *Frattini's argument* is a very useful result (cf. [65, Theorem 1.3.7]).

**Lemma 1.1.** *Let $G$ be a finite group, $H \trianglelefteq G$ and $S \in \mathrm{Syl}_p(H)$. Then $G = N_G(S)H$.*

An *elementary abelian $p$-subgroup* of $G$ is an abelian $p$-subgroup $E$ of $G$ of exponent $p$. That is, $E \cong C_p \times \cdots \times C_p$, where $C_p$ denotes a cyclic group of order $p$. Switching to additive notation, $E \cong \mathbb{Z}/p \oplus \cdots \oplus \mathbb{Z}/p$, or $E \cong \mathbb{F}_p^r$ as an $\mathbb{F}_p$-vector space, where $r = \dim_{\mathbb{F}_p}(E) = \log_p |E|$ is the *rank* of $E$. The *$p$-rank* of $G$ is the *rank* of $S \in \mathrm{Syl}_p(G)$, namely, it is the maximum of the ranks of the elementary abelian $p$-subgroups of $G$ (or equivalently $S$). A *maximal* elementary abelian $p$-subgroup is an elementary abelian $p$-subgroup which is not properly contained in another elementary abelian $p$-subgroup of $G$. Such a $p$-subgroup need not have maximal rank. For instance, if $p$ is odd, the *wreath product* $C_p \wr C_p$ has ($p$-)rank $p$ and has a maximal elementary abelian ($p$-)subgroup of rank 2.

Let $S$ be a nontrivial finite $p$-group. Then $Z(S) \neq 1$ and $S$ is nilpotent of class at most $\log_p(|S|) - 1$. We say that $S$ has *maximal nilpotency class* if $S$ is nilpotent of class exactly $\log_p |S| - 1$. For any $S$, the quotient group $S/\Phi(S)$ is elementary abelian of rank equal to the minimum number of generators of $S$ (cf. [81, III.3.15 Burnsidescher Basissatz]).

An *extraspecial $p$-group* is a $p$-group $G = p^{1+2n}$ of order $p^{1+2n}$ and nilpotency class 2 with $\Phi(G) = G' = Z(G) \cong C_p$. They have been classified, see for instance [65, Sect. 5.5]. Namely, they are central products of extraspecial groups of order $p^3$, denoted $p_+^{1+2}$ and $p_-^{1+2}$, of exponent $p$ and $p^2$ respectively if $p$ is odd, and, for $p = 2$, we set $2_+^{1+2}$ for the dihedral group of order 8 and $2_-^{1+2}$ for the quaternion group of order 8. Thus $p_+^{1+2n}$ is the central product of $n$ copies of $p_+^{1+2}$, and $p_-^{1+2n}$ is the central product of one copy of $p_-^{1+2}$ and $(n-1)$ copies of $p_+^{1+2}$, for all primes $p$ and all positive integers $n$.

A *$p$-local subgroup* of $G$ is the normaliser of a nontrivial $p$-subgroup of $G$. The $p$-local subgroups of a finite group are key in studying the $p$-fusion in $G$, as can be seen in *Alperin's fusion theorem* [63, Theorem 3.5]. There are several versions of this theorem, depending on which family of subgroups we consider. Let us record the following terminology related to $p$-fusion which will be needed later. If $S \in \mathrm{Syl}_p(G)$ and $x, y \in S$, we say that $x$ and $y$ *fuse in $G$* if $x$ and $y$ are not conjugate in $S$ but are conjugate in $G$; and similarly for $p$-subgroups $Q, R$ of $S$, we say that they fuse in $G$ if they are not conjugate in $S$ but are conjugate in $G$ (cf. [65, Chap. 7]). A *$p$-centric subgroup* is a $p$-subgroup $Q$ of $G$ such that $Z(Q) \in \mathrm{Syl}_p(C_G(Q))$. A $p$-centric subgroup $Q$ of $G$ is *centric*, or *self-centralising* if $Z(Q) = C_G(Q)$. A *$p$-radical subgroup* of $G$ is a $p$-subgroup $Q$ of $G$ such that $\mathrm{Out}_G(Q)$ has no nontrivial normal $p$-subgroup. We call $Q$ *essential* if $C_G(Q) = Z(Q)$ and $\mathrm{Out}_G(Q)$ has a strongly $p$-embedded subgroup.

## 1.3   Modular Representation Theory

**Definition 1.1.** A *p-modular system* is a triple $(F, \mathcal{O}, k)$ where $\mathcal{O}$ is a complete rank one discrete valuation ring of characteristic 0 with unique maximal ideal $\pi$, $F$ its field of fractions and $k = \mathcal{O}/\pi\mathcal{O}$ the residue field of $\mathcal{O}$, which has characteristic $p$. We say that $(F, \mathcal{O}, k)$ is *large enough for G* if $F$ contains a primitive $m$-th root of unity where $m$ is the exponent of $G$ and if $k = \bar{k}$ is algebraically closed.

The *modular representations* of $G$ are the $RG$-modules where $R = \mathcal{O}$ or $k$ in a *p*-modular system. We will always assume that our *p*-modular systems are large enough and that our modules are finitely generated $R$-free left $RG$-modules. We refer the reader to [1, 10, 11, 50, 51, 60, 85, 118] for more details on the subject.

We write $R$ for the ring of coefficients and for the trivial $RG$-module. The tensor product symbol $\otimes$ denotes $\otimes_R$, the tensor product in the category of free $R$-modules. Given two $RG$-modules $M, N$, we write $M|N$ if $M$ is isomorphic to a direct summand of $N$. The set $\mathrm{Hom}_R(M, N)$ of $R$-linear homomorphisms $M \to N$ and the tensor product $M \otimes N$ are $RG$-modules with respect to the $G$-actions

$$(g \cdot \varphi)(m) = g\varphi(g^{-1}m) \quad \text{and} \quad g(m \otimes n) = gm \otimes gn \tag{1.1}$$

for all $m \in M$, $n \in N$, $g \in G$ and $\varphi \in \mathrm{Hom}_R(M, N)$.

Let $M$ be an $RG$-module. The $R$-*dual*, or simply the *dual*, of $M$ is the $RG$-module $M^* = \mathrm{Hom}_R(M, R)$, the set of $R$-linear transformations $M \to R$. This is an $RG$-module with respect to the $G$-action

$$(g \cdot \varphi)(m) = \varphi(g^{-1}m) \quad \text{for all } m \in M, \ g \in G \text{ and } \varphi \in \mathrm{Hom}_R(M, R).$$

Unless otherwise specified, the maps act on the left of the modules and composition of maps reads right to left, i.e. the composition

$$A \xrightarrow{\varphi} B \xrightarrow{\psi} C \quad \text{is written} \quad \psi\varphi : A \longrightarrow C.$$

Tensor products over $R$ and $R$-duals of $RG$-modules will be used often in the sequel, and in particular the following isomorphism.

**Lemma 1.2.** ([10, Lemma 2.1.1]) *Let $M, N$ be $RG$-modules. Then*

$$\mathrm{Hom}_R(M, N) \cong M^* \otimes N \quad \text{as } RG-\text{modules}.$$

We now introduce two very special classes of modules, and also the two main categories of modules, which we use in the study of endotrivial modules, and homological algebra more broadly.

**Definition 1.2.** An $RG$-module $P$ is *projective* if and only if, every surjective $RG$-homomorphism $\pi : M \twoheadrightarrow P$ splits, i.e. there exists a $kG$-homomorphism $\sigma :$

$P \to M$ such that $\pi\sigma = \mathrm{id}_P$. An $RG$-module $J$ is *injective* if every injective $RG$-homomorphism $\phi \in \mathrm{Hom}_{RG}(J, M)$ has a *retraction*, that is, there exists a $\psi \in \mathrm{Hom}_{RG}(M, J)$ such that $\psi\phi = \mathrm{id}_J$.

We refer to [11, Vol I, Sect. 1.5] for the equivalent characterisations of projective and injective modules.

Since $RG$ is a symmetric algebra, the rank one free $RG$-module $RG$ is selfdual, $RG \cong RG^*$, and it follows that $P$ is projective if and only if $P$ is injective. If $G$ is a finite $p$-group, then any projective module is free since $RG$ is a local ring.

We write $\mathrm{mod}(RG)$ for the category of finitely generated $RG$-modules, whose objects are the finitely generated $RG$-modules, and the morphisms $\mathrm{Hom}_{RG}(M, N)$ between two $RG$-modules are the $RG$-homomorphisms between $RG$-modules.

We write $\mathrm{stmod}(RG)$ for the *stable module category*, whose objects are the finitely generated $RG$-modules, and the morphisms $\underline{\mathrm{Hom}}_{RG}(M, N)$ between two $RG$-modules are the equivalence classes of $RG$-homomorphisms with respect to the equivalence relation: *$f \sim g$ for $f, g \in \mathrm{Hom}_{RG}(M, N)$ if and only if there exists a projective $RG$-module $P$ and $RG$-homomorphisms $s \in \mathrm{Hom}_{RG}(M, Q)$ and $t \in \mathrm{Hom}_{RG}(Q, N)$ such that $f - g = ts$, i.e. such that the diagram*

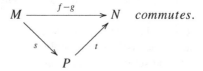

In particular, in $\mathrm{stmod}(RG)$, projective modules are isomorphic to 0 since the identity map of a projective module $P$ trivially factors through a projective module, i.e. $\mathrm{id}_P \sim 0_P$. (cf. [11, Vol I, Sect. 2.1]).

### 1.3.1  Restriction, Inductions and Inflation

Given a group homomorphism $f : G \to H$, we write $\mathrm{Res}_f : \mathrm{mod}(RH) \to \mathrm{mod}(RG)$ for the *restriction along $f$*, which turns an $RG$-module $M$ into an $RH$-module by the rule $hm = f(h)m$ for $h \in H$ and $m \in M$.

If $f$ is the inclusion of a subgroup $H$, we write $\mathrm{Res}_H^G M$ or $M{\downarrow}_H^G$ instead of $\mathrm{Res}_f M$ for an $RG$-module $M$.

If $f$ is surjective, i.e. $H \cong G/N$ for the normal subgroup $N = \ker f$ of $G$, then $\mathrm{Res}_f$ is the *inflation* from $H$ to $G$, denoted by $\mathrm{Inf}_H^G$.

The *induction* of an $RH$-module $M$ to an overgroup $G$ of $H$ is the $RG$-module $RG \otimes_{RH} M$ equipped with the $G$-action $g(x \otimes m) = (gx) \otimes m$, and we write either $\mathrm{Ind}_H^G M$ or $M{\uparrow}_H^G$. As a free $R$-module, $M{\uparrow}_H^G \cong \bigoplus_{x \in [G/H]} {}^x M$, where $[G/H]$ is a set of coset representatives of $H$ in $G$ and ${}^x M$ (sometimes written $x \otimes M$ in the literature) is the *conjugate* $R[^xH]$-*module* of $M$, where $^xh(x \otimes m) = x \otimes hm$ for all $^xh \in {}^xH$ and $x \otimes m \in {}^xM$.

A less well known kind of induction is the *tensor induction* $\mathrm{Ten}_H^G M$, sometimes written $M{\uparrow}_H^{\otimes G}$ (cf. [11, Vol I, Sect. 3.15]).

**Definition 1.3.** Let $H$ be a subgroup of a finite group $G$ and fix a set $[G/H]$ of coset representatives. Let $M$ be an $RH$-module. The *tensor induction* $\mathrm{Ten}_H^G M$ of $M$ is the $RG$-module with underlying free $R$-module $\bigotimes\limits_{s \in [G/H]} {}^sM$. To define the $G$-action on $\mathrm{Ten}_H^G M$, we use the permutation action of $G$ on $[G/H]$. That is, given $g \in G$ and $s \in [G/H]$, put $gs = \sigma_g(s)h_{g,s}$ for some $\sigma_g(s) \in [G/H]$ and $h_{g,s} \in H$. Then,

$$g \left( \bigotimes_{s \in [G/H]} s \otimes x_s \right) = \bigotimes_{s \in [G/H]} \sigma_g(s) \otimes h_{g,s}x_s = \bigotimes_{s \in [G/H]} s \otimes h_{g,\sigma_g^{-1}(s)}x_{\sigma_g^{-1}(s)},$$

for all $g \in G$ and all $\bigotimes\limits_{s \in [G/H]} s \otimes x_s \in \mathrm{Ten}_H^G M$.

In particular, $\mathrm{Ten}_H^G R = R$ for any $H \leq G$, and the tensor induction of a projective module is not projective in general. Note also that $\mathrm{rank}_R(\mathrm{Ten}_H^G M) = (\mathrm{rank}_R M)^{|G:H|}$, whereas $\mathrm{rank}_R(\mathrm{Ind}_H^G M) = |G : H| \mathrm{rank}_R M$.

A sensible way to regard $\mathrm{Ten}_H^G M$ is to consider it first as an $R[H \wr \mathfrak{S}_{|G:H|}]$-module, where $\mathfrak{S}_{|G:H|}$ is the symmetric group permuting the cosets $G/H$, and then take the restriction along the inclusion $G \hookrightarrow H \wr \mathfrak{S}_{|G:H|}$.

We end this section with a useful result on tensor induction ([11, Vol I, Proposition 3.15.2]).

**Proposition 1.1.** *Let $H \leq G$ and $M, N$ be two $RH$-modules.*

1. $\mathrm{Ten}_H^G(M \otimes N) \cong \mathrm{Ten}_H^G M \otimes \mathrm{Ten}_H^G N$.
2. *If $H \leq K \leq G$, then $\mathrm{Ten}_K^G\left(\mathrm{Ten}_H^K M\right) \cong \mathrm{Ten}_H^G M$.*
3. $\mathrm{Ten}_H^G(M \oplus N) \cong \mathrm{Ten}_H^G M \oplus \mathrm{Ten}_H^G N \oplus X$, *where $X$ is a direct sum of modules induced from proper subgroups of $G$ containing the intersection of the conjugates of $H$.*

### 1.3.2 Blocks

The group algebra $RG$ decomposes as a direct sum of indecomposable subalgebras called the *blocks* of $RG$, possibly $RG$ is a single block. The blocks are determined by the decomposition of $1_{RG} = e_0 + \cdots + e_r$ into a sum of central idempotents $e_j^2 = e_j \in Z(RG)$ which are primitive and pairwise orthogonal. *Primitive* means that an $e_j$ cannot be written as $e_j' + e_j'' = e_j$ for nonzero idempotents $e_j', e_j'' \in Z(RG)$ with $e_j'e_j'' = 0$, and *pairwise orthogonal* means that $e_ie_j = \delta_{ij}$ for all $0 \leq i, j \leq r$.

We call each such $e_j$ a *block idempotent*, and write $B_j = RGe_j$ for the block algebra of $RG$ with multiplicative identity $e_j$, and annihilated by multiplication by $e_i$ for all $i \neq j$. Correspondingly, an $RG$-module splits as the direct sum of modules $M = M_1 \oplus \cdots \oplus M_r$, where $M_j = e_j M$ *lies in the block* $B_j$, for all $j$. That is, each $M_j$ is a submodule of $M$ on which the idempotent $e_j$ of $B_j$ acts as the identity, i.e. $e_j m = m$ for all $m \in M_j$, while $e_i m = 0$ for all $i \neq j$ and all $m \in M_j$. In particular, if $M$ is indecomposable, then $M$ must belong to some $B_j$. The *principal block* is the block $B_0$ containing the trivial $RG$-module $R$. We refer to [1, 50, 84, 118] for the theory of blocks and their numerous invariants. Let us only mention the following facts without proof (cf. [1, Sects. 3 and 4]).

**Theorem 1.1.**   *1.  If $G$ is a p-group, then the group algebra $RG$ is indecomposable as an $RG$-module, i.e. there is a unique block. Furthermore, $R$ is the unique simple $RG$-module.*
2.  *If $G$ is any finite group and $M$ an $RG$-module, then $M$ is indecomposable if and only if $\operatorname{End}_{RG} M$ is a local ring.*

The additive and multiplicative structures on $\operatorname{End}_{RG} M$ are the pointwise addition and the composition of maps respectively.

### 1.3.3   The Krull–Schmidt Theorem

Because the group algebra $RG$ is not semisimple, we cannot decompose every module into a direct sum of simple modules. However, each module has the *unique decomposition property*, which essentially reduces the study of the representations of $G$ over $R$ to that of the indecomposable $RG$-modules. We say that a ring $\Lambda$ has the unique decomposition property if for every finitely generated $\Lambda$-module $M$, the following hold (cf. [60, Sect. I.11]):

1.  $M$ is a direct sum of finitely many indecomposable $\Lambda$-modules.
2.  If $M \cong \oplus_{i=1}^m V_i \cong \oplus_{j=1}^n W_j$, where each $V_i$, $W_j$ is indecomposable, then $n = m$ and $V_i \cong W_i$ after a suitable rearrangement.

A $\Lambda$-module is *indecomposable* if it is nonzero and cannot be expressed as the direct sum of two nonzero $\Lambda$-submodules. The main result is the Krull–Schmidt theorem [118, Theorem I.4.4].

**Theorem 1.2.**  *Let $A$ be an $\mathcal{O}$-algebra. Then $A$ has the unique decomposition property. That is, for every finitely generated $A$-module $M$:*

1.  *there exists a decomposition $M \cong \oplus_{i=1}^n V_i$ into a finite direct sum of indecomposable $A$-modules $V_i$, and*
2.  *for any decompositions $M \cong \oplus_{i=1}^n V_i \cong \oplus_{j=1}^m W_j$ of $M$ into direct sums of indecomposable $A$-modules, $m = n$ and $V_i \cong W_i$ after a suitable rearrangement of the summands.*

### 1.3.4 Mackey's Formula

Mackey's formula is also known as the Subgroup theorem (cf. [50, (10.13)]).

**Theorem 1.3.** *Let* $H, K \leq G$ *and* $M$ *be an* $RH$-*module. Then*

$$M\uparrow_H^G\downarrow_K^G \cong \bigoplus_{x \in [K \backslash G / H]} {}^xM\downarrow_{{}^xH \cap K}^{{}^xH}\uparrow_{{}^xH \cap K}^K,$$

*where* ${}^xM$ *is the conjugate* $R[{}^xH]$-*module of* $M$ *by* $x$.

In particular, if $K = H \trianglelefteq G$, then Mackey's formula gives

$$M\uparrow_H^G\downarrow_H^G \cong \bigoplus_{x \in [G/H]} {}^xM \quad \text{is a sum of } conjugate \, RH\text{-}modules.$$

Note that, since $H$ is normal in $G$, we can make ${}^xM$ into an $RH$-module with respect to the action $h({}^xm) = {}^x((h^x)m)$. Thus $M \not\cong {}^xM$ as $RH$-modules in general.

**Definition 1.4.** Let $H \trianglelefteq G$ and $M$ be an $RH$-module. The *inertia group* of $M$ is the subgroup

$$I(M) = \{x \in G \, : \, {}^xM \cong M \text{ as } RH\text{-modules } \} \quad \text{of } G.$$

Note that $I(M)$ is indeed a group and that $HC_G(H) \leq I(M)$. The following is [11, Vol I, Sect. 3.13].

**Proposition 1.2.** *Let* $H \trianglelefteq G$ *and* $M$ *be an* $RH$-*module with inertia group* $I(M)$. *Suppose that* $M\uparrow_H^{I(M)} \cong V_1 \oplus \cdots \oplus V_n$ *with* $V_i$ *indecomposable for each* $i$. *Then* $V_i\uparrow_{I(M)}^G$ *is indecomposable and* $V_i\uparrow_{I(M)}^G \cong V_j\uparrow_{I(M)}^G$ *if and only if* $j = i$, *for all* $i, j$.

There is a version of Mackey's formula which applies to tensor induction [11, Vol I, Proposition 3.15.2].

**Proposition 1.3.** *Let* $H, K \leq G$ *and* $M$ *be an* $RH$-*module. Then*

$$\mathrm{Res}_K^G \, \mathrm{Ten}_H^G \, M \cong \bigotimes_{x \in [K \backslash G / H]} \mathrm{Ten}_{K \cap {}^xH}^K \, \mathrm{Res}_{{}^xH \cap H}^{{}^xH}({}^xM).$$

## 1.4 Homological Algebra

Homological algebra methods are ubiquitous in representation theory, and especially in the study of endotrivial modules, as we shall see. In this section we outline the notions and results of homological algebra that we need. For convenience, we introduce the notation $\hookrightarrow$ and $\twoheadrightarrow$ to denote an inclusion and a surjection, respectively.

### 1.4.1   Exact Sequences and Projective Resolutions

The conventions in homological algebra vary between authors. We will mainly follow [11, 46], and so adopt the following notation and terminology.

A *complex* in the category mod($RG$) is a sequence of $RG$-modules and $RG$-homomorphisms

$$(A_*, \alpha_*) = \quad \cdots A_2 \xrightarrow{\alpha_2} A_1 \xrightarrow{\alpha_1} A_0 \xrightarrow{\alpha_0} A_{-1} \cdots$$

such that $\operatorname{im}(\alpha_{j+1}) \subseteq \ker(\alpha_j)$ for all $j$. We call $\alpha_j$ a *differential*.

A complex $(A_*, \alpha_*)$ *splits* if, for each $j$, there exists an $RG$-homomorphism $\beta_j : A_{j-1} \to A_j$ such that $\alpha_j \beta_j \alpha_j = \alpha_j$.

A complex $(A_*, \alpha_*)$ is an *exact sequence* if $\ker(\alpha_j) = \operatorname{im}(\alpha_{j+1})$ for all $j$. We call an exact sequence *short* if $A_j = 0$ for all but three consecutive indices. Note that an exact sequence $(A_*, \alpha_*)$ splits if $A_j \cong \ker(\alpha_j) \oplus \operatorname{im}(\alpha_j)$ for all $j$. In particular, a short exact sequence $0 \to A_2 \xrightarrow{\alpha_2} A_1 \xrightarrow{\alpha_1} A_0 \to 0$ splits if and only if $A_1 \cong A_2 \oplus A_0$, by [26, Lemma 6.12].

A *morphism* $f_* : (A_*, \alpha_*) \to (B_*, \beta_*)$ between two complexes is a family of $RG$-homomorphisms $f_j : A_j \to B_j$ such that $f_j \alpha_{j+1} = \beta_{j+1} f_{j+1}$ for all $j$, i.e. the diagram

$$
\begin{array}{ccc}
\cdots \longrightarrow A_{j+1} & \xrightarrow{\alpha_{j+1}} & A_j \longrightarrow \cdots \\
\downarrow{\scriptstyle f_{j+1}} & & \downarrow{\scriptstyle f_j} \\
\cdots \longrightarrow B_{j+1} & \xrightarrow{\beta_{j+1}} & B_j \longrightarrow \cdots
\end{array}
\qquad \text{commutes.}
$$

A *projective resolution* of an $RG$-module $M$ is a long exact sequence

$$(P_*, \partial_*) = \quad \cdots P_2 \xrightarrow{\partial_2} P_1 \xrightarrow{\partial_1} P_0$$

of $RG$-modules with $P_j$ projective for all $j$ and such that $P_0/\operatorname{im}(\partial_1) \cong M$. In particular, we have an exact sequence

$$P_0 \longrightarrow M \longrightarrow 0 , \quad \text{called a } projective\ cover\ of\ M.$$

We know that every $RG$-module has a projective resolution, and that any two projective resolutions are *homotopy equivalent* ([11, Vol I, Sects. 2.3 and 2.4]). That is, if $(P'_*, \partial'_*)$ is another projective resolution of $M$, then there exist morphisms of complexes $f_* : (P_*, \partial_*) \to (P'_*, \partial'_*)$ and $f'_* : (P'_*, \partial'_*) \to (P_*, \partial_*)$, and two collections of $RG$-homomorphisms $\{h_n \in \operatorname{Hom}_{RG}(P_n, P_{n+1}) \mid n \in \mathbb{Z}\}$ and $\{h'_n \in \operatorname{Hom}_{RG}(P'_n, P'_{n+1}) \mid n \in \mathbb{Z}\}$, such that $\partial_{n+1} h_n + h_{n-1} \partial_n = f'_n f_n - \operatorname{id}_{P_n}$ and

$\partial'_{n+1} h'_n + h'_{n-1} \partial'_n = f_n f'_n - \mathrm{id}_{P'_n}$ for all $n$. (The reader not familiar with the notion of homotopy equivalence is advised to draw a diagram of the situation.)

By symmetry, every $RG$-module $M$ has an *injective resolution*, that is, an exact sequence $(P_*, \partial_*) = P_0 \xrightarrow{\partial_0} P_{-1} \xrightarrow{\partial_{-1}} P_{-2} \cdots$ of injective $RG$-modules with $M \cong \ker(\partial_0)$.

Hence a *complete resolution* of $M$ is the *splice*

$$\cdots P_2 \xrightarrow{\partial_2} P_1 \xrightarrow{\partial_1} P_0 \xrightarrow{\partial_0} P_{-1} \xrightarrow{\partial_{-1}} P_{-2} \cdots$$

of a projective and an injective resolution of $M$.

As a consequence, the (co-)kernels in any two complete resolutions of an $RG$-module $M$ are isomorphic in $\mathrm{stmod}(RG)$, and they do not depend on the choice of such resolutions.

**Definition 1.5.** Let $M$ be an $RG$-module and $(P_*, \partial_*)$ a complete resolution of $M$. We define the *n-th syzygy* $\Omega_G^n(M) = \ker\left(\partial_{n-1} : P_{n-1} \to P_{n-2}\right)$ of $M$ for all $n \in \mathbb{Z}$. If $n = 1$, we write $\Omega_G(M)$ instead of $\Omega_G^1(M)$.

In particular, for every $RG$-module $M$, there is a short exact sequence

$$0 \longrightarrow \Omega_G(M) \longrightarrow P_0 \longrightarrow M \longrightarrow 0$$

where $P_0 \to M \to 0$ is a projective cover of $M$.

### 1.4.2 Relative Projectivity

An $RG$-module $M$ is *projective relative* to a subgroup $H$ of $G$ if $M \mid V \uparrow_H^G$ for some $RH$-module $V$. Any $RG$-module is projective relative to a subgroup containing a Sylow $p$-subgroup of $G$ and the projective $RG$-modules are precisely those which are projective relative to the trivial group ([1, Sect. 9]). So relative projectivity generalises projectivity. In particular, given any subgroup $H$ of $G$, every $RG$-module has a relatively $H$-projective resolution which is unique up to homotopy (cf. [11, Vol I, Sect. 3.9]).

An important construction in homological algebra is the relative trace (cf. [11, Vol I, Sect. 3.6]).

**Definition 1.6.** Let $H$ be a subgroup of $G$ and $A$ a $G$-algebra over $R$. The *(relative) trace map* is the $R$-linear map

$$\mathrm{tr}_H^G : A^H \to A^G, \mathrm{tr}_H^G(a) = \sum_{g \in [G/H]} {}^g a \quad \text{for all } a \in A,$$

where $[G/H]$ is a set of coset representatives of $H$ in $G$.

In the particular case when $M$, $N$ are two $RG$-modules and $A = \mathrm{Hom}_R(M, N)$, then

$$\mathrm{tr}_H^G \;:\; \mathrm{Hom}_{RH}(M, N) \to \mathrm{Hom}_{RG}(M, N)\,, \quad \mathrm{tr}_H^G(\phi) = \sum_{g \in [G/H]} g\phi,$$

where $\mathrm{tr}_H^G(\phi)(m) = \sum_{g \in [G/H]} g\phi(g^{-1}m)$ for $\phi \in \mathrm{Hom}_{RH}(M, N)$ and $m \in M$.

The following is an adaptation of [11, Vol I, Proposition 3.6.4].

**Proposition 1.4.** (Higman's criterion) *Let $H$ be a subgroup of $G$ and $M$ be an $RG$-module. The following are equivalent.*

1. *$M \mid \mathrm{Ind}_H^G \mathrm{Res}_H^G M$ as an $RG$-module.*
2. *$M \mid \mathrm{Ind}_H^G V$ for some $RH$-module $V$, i.e. $M$ is projective relative to $H$.*
3. *There exists a $\varphi \in \mathrm{End}_{RH} M$ such that $\mathrm{tr}_H^G(\varphi) = \mathrm{id}_M$.*
4. *Given a commutative diagram*

   *of $RG$ −modules, where*

   *$\beta$ and $\pi$ are $RG$-homomorphisms and where $\alpha$ is an $RH$-homomorphism, there exists an $\hat{\alpha} \in \mathrm{Hom}_{RG}(M, A)$ such that $\pi\hat{\alpha} = \beta$.*
5. *Given a surjective $RG$-homomorphism $\pi \in \mathrm{Hom}_{RG}(A, M)$, if there exists an $\alpha \in \mathrm{Hom}_{RH}(M, A)$ such that $\pi\alpha = \mathrm{id}_M$, then there exists an $\hat{\alpha} \in \mathrm{Hom}_{RG}(M, A)$ such that $\pi\hat{\alpha} = \mathrm{id}_M$.*

The maps $\alpha$, $\hat{\alpha}$ in Proposition 1.4 Part 5 are $H$-*sections*, and we say that $\pi$ is $H$-split.
   A *pull back* is the completion of a diagram of objects and morphisms

$$\begin{array}{ccc} & & A \\ & & \downarrow{\scriptstyle\psi} \\ B & \xrightarrow{\phi} & C \end{array}$$

in a given category $\mathscr{C}$ into a commutative diagram

$$\begin{array}{ccc} X & \xrightarrow{\hat{\phi}} & A \\ {\scriptstyle\hat{\psi}}\downarrow & & \downarrow{\scriptstyle\psi} \\ B & \xrightarrow{\phi} & C \end{array}$$

in $\mathscr{C}$. One can show that $\ker(\phi) = \ker(\hat{\phi})$ and that if $\phi$ is $H$-split, then so is $\hat{\phi}$. Similarly for $\psi$.

**Lemma 1.3.** (Relative Schanuel's lemma) *Given a diagram of exact sequences of RG-modules:*

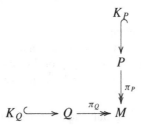

*where $P$ and $Q$ are projective relative to a subgroup $H$ of $G$ and $\pi_P, \pi_Q$ are $H$-split, then $K_Q \oplus P \cong K_P \oplus Q$ as RG-modules.*

*Proof.* Construct the pull back of this diagram

$$
\begin{array}{ccc}
K_P & \xrightarrow{\ =\ } & K_P \\
\downarrow & & \downarrow \\
K_Q \hookrightarrow X & \xrightarrow{\hat{\pi}_P} & P \\
\Big\downarrow{=} \quad \hat{\pi}_Q\Big\downarrow & & \Big\downarrow{\pi_P} \\
K_Q \hookrightarrow Q & \xrightarrow{\pi_Q} & M
\end{array}
$$

where we have identified the kernels. Because $\pi_P, \pi_Q$ are $H$-split, so too are $\hat{\pi}_P$ and $\hat{\pi}_Q$. Since they both are surjective, Proposition 1.4 says that they are split as $RG$-homomorphisms and therefore

$$K_P \oplus Q \cong X \cong K_Q \oplus P.$$   □

**Definition 1.7.** Let $H$ be a subgroup of $G$ and $M$ an $RG$-module. The kernel

$$\Omega_{G/H}(M) = \ker(\ P \longrightarrow M\ )$$

of a relatively $H$-projective cover of $M$ is the *$H$-relative syzygy* of $M$. More generally, given a minimal relatively $H$-projective resolution $\cdots P_1 \xrightarrow{\ \delta_0\ } P_0 \longrightarrow M$ of $M$ the *$H$-relative $n$-th syzygy* of $M$ is the kernel $\Omega_{G/H}^n(M) = \ker(\delta_n)$. If $n = 1$, we write $\Omega_{G/H}^1(M) = \Omega_{G/H}(M)$. A *relative syzygy* is an $H$-relative syzygy for some subgroup $H$ of $G$.

We take *minimal* relative projective covers, so that the relative syzygies are indecomposable modules obtained as the (co-)kernels in the resolution.

In particular, if $G$ is a finite $p$-group and $M = R$, then

$$\Omega_{G/H}(R) = \ker(\ R[G/H] \xrightarrow{\ \varepsilon\ } R\ )$$

is the kernel of the augmentation map $\varepsilon$ sending each coset $gH$ to 1.

The concept of relative projectivity can be extended to relative projectivity with respect to a family of subrings of $RG$ [119], or $RG$-modules [26]. In the context of endotrivial modules, we will also need relative projectivity with respect to some $RG$-module.

**Definition 1.8.** Let $G$ be a finite group and $V$ an $RG$-module. An $RG$-module $M$ is *relatively $V$-projective* if there exists an $RG$-module $N$ such that $M \mid V \otimes N$.

A short exact sequence of $RG$-modules

$$0 \longrightarrow M_1 \longrightarrow M_2 \longrightarrow M_3 \longrightarrow 0$$

is *$V$-split* if the short exact sequence obtained by applying the functor $V \otimes -$,

$$0 \longrightarrow V \otimes M_1 \longrightarrow V \otimes M_2 \longrightarrow V \otimes M_3 \longrightarrow 0$$

splits.

Note that "$R$-projective" coincides with "projective" for projectivity relative to the trivial module.

### 1.4.3  Vertices, Sources of Modules and Green's Correspondence

Given an indecomposable $RG$-module $M$ there exist a $p$-subgroup $Q$ of $G$ of minimal order and an indecomposable $RQ$-module $V$ such that $M \mid V{\uparrow}_Q^G$. Such a subgroup $Q$ is unique up to conjugation and $V$ is unique up to isomorphism and conjugation. We call $Q$ a *vertex* of $M$ and $V$ a *source* of $M$. In particular, an $RG$-module is called *trivial source* if its source is the trivial module $R$. We will discuss these in greater detail in Sect. 1.5.

*Green's correspondence* is a fundamental result in the modular representation theory of finite groups which provides a $1-1$ correspondence between the isomorphism classes of indecomposable $RG$-modules with given vertex $Q$, a $p$-subgroup of $G$, and the indecomposable $RH$-modules with vertex $Q$, for any subgroup $H$ of $G$ containing $N_G(Q)$, provided the $p'$-part $|G|_{p'}$ of $|G|$ is invertible in $R$ (cf. [11, Vol I, Sect. 3.12]). Given $Q$, $H$ and $R$ as above, define the sets

$$\mathfrak{x} = \{{}^g Q \cap Q \mid g \in G - H\} \quad \text{and} \quad \mathfrak{y} = \{{}^g Q \cap H \mid g \in G - H\}$$

of $p$-subgroups of $G$.

**Theorem 1.4.** *Let $Q$ be a $p$-subgroup of $G$ and $H$ a subgroup of $G$ containing $N_G(Q)$. Write $\mathfrak{x}$ and $\mathfrak{y}$ as above. There exists a $1 - 1$ correspondence between the isomorphism classes of indecomposable $RG$-modules with vertex $Q$ and the indecomposable $RH$-modules with vertex $Q$, given as follows.*

1. *Let $V$ be an indecomposable $RG$-module with vertex $Q$. Write $V\downarrow_H^G = \bigoplus_{i \in I} V_i$ as a sum of indecomposable $RH$-modules. Then there exists a unique $i$ such that $V_i$ has vertex $Q$, and all the other indecomposable summands have vertex in $\mathfrak{y}$. We write $V_i = \Gamma_H(V)$ for any such pairing of $RG$- and $RH$-modules $V$ and $V_i$.*
2. *Let $U$ be an indecomposable $RH$-module with vertex $Q$. Write $U\uparrow_H^G = \bigoplus_{j \in J} U_j$ as a sum of indecomposable $RG$-modules. Then there exists a unique $j$ such that $U_j$ has vertex $Q$, and all the other indecomposable summands have vertex in $\mathfrak{x}$. We write $U_j = \Gamma^G(U)$ for any such pairing of $RH$- and $RG$-modules $U$ and $U_j$.*
3. *We have $\Gamma_H(\Gamma^G(U)) \cong U$ and $\Gamma^G(\Gamma_H(V)) \cong V$.*

The modules $\Gamma_H(V)$ and $\Gamma^G(U)$ in Theorem 1.4 are called the $RH$- and $RG$-*Green correspondents* of $V$ and $U$, respectively . In the literature, they are often written as $f(V)$ instead of $\Gamma_H(V)$ and $g(U)$ instead of $\Gamma^G(U)$ (cf. [11, Vol I, Sect. 3.12]), however, to avoid possible confusion with some $kG$-homomorphism $f$, or some group element $g \in H$, we prefer to use the notation $\Gamma_H(-)$ and $\Gamma^G(-)$, respectively.

This theorem raises the question of detecting the Green correspondent of an $RG$- or $RH$-module. The Burry–Carlson–Puig theorem [11, Vol I, Theorem 3.12.3] provides some hints.

**Theorem 1.5.** *Let $Q$ be a $p$-subgroup of $G$, let $H$ be a subgroup of $G$ containing $N_G(Q)$, and let $V$ be an indecomposable $RG$-module. Suppose that $V\downarrow_H^G$ has an indecomposable direct summand $U$ with vertex $Q$. Then $V$ has vertex $Q$ and $U \cong \Gamma_H(V)$.*

A special case of Green's correspondence occurs for groups with a strongly $p$-embedded subgroup, which happens for instance if a Sylow $p$-subgroup is a trivial intersection subgroup (cf. [1, Sect. 10]).

### 1.4.4  Green's Indecomposability Criterion

Another key result due to J. A. Green is the following criterion for indecomposability [11, Vol I, Theorem 3.13.3].

**Theorem 1.6.** *Suppose that $N$ is a normal subgroup of $G$ with factor group $G/N$ of $p$-power order, and let $V$ be an indecomposable $RN$-module. Then $V\uparrow_N^G$ is also indecomposable.*

Building on this result, since we assume that $p$ divides $|G|$, any projective $RG$-module has dimension divisible by $|G|_p$, the $p$-part of $|G|$, and so, if $G$ is a $p$-group, then $RG$ is indecomposable. It is in fact the only indecomposable projective $RG$-module.

## 1.5 ($p$-)Permutation Modules

Forgetting the $G$-action, an $RG$-module $M$ is a free $R$-module of finite rank. As such, $M$ has an $R$-basis, that is, a subset $X$ of $M$ such that each $m \in M$ can be uniquely written as an $R$-linear combination $m = \sum_{x \in X} \lambda_x x$ for coefficients $\lambda_x \in R$.

Recall that a $G$-set is a set $X$ equipped with an action $G \times X \to X$ of $G$, written $(g, x) \mapsto gx$, such that $g(hx) = (gh)x$ and $1x = x$, for all $g, h \in G$ and all $x \in X$.

**Definition 1.9.** Let $M$ be an $RG$-module. We call $M$ a *permutation $RG$-module* if $M$ has an $R$-basis which is also a $G$-set. We call $M$ a *$p$-permutation module* if $M{\downarrow}_S^G$ is a permutation $RS$-module for a Sylow $p$-subgroup $S$ of $G$.

For instance, the regular module $RG$ and the trivial module $R$ are permutation modules with respect to the bases $G$ (regarded as a set) and $\{\sum_{g \in G} g\}$, respectively.

Note that permutation modules are $p$-permutation too. References [60, Sect. IX.3], [16, Chap. 12] and [21] give the main properties of permutation modules, and [23] those of $p$-permutation modules.

There is a functor from the category of finite left $G$-sets to the category of permutation modules which assigns to each $G$-set $X$ the permutation $RG$-module $RX$, defined to be the free $R$-module with basis $X$. Such a module is *transitive* if $G$ acts transitively on $X$, i.e. $X$ is a single $G$-orbit. Thus $RX \cong R[G/G_x]$, where $G_x$ is the stabiliser of some $x \in X$ (isotropy group of $x$).

Let us collect (without proof) some of the main facts ([60, Chap. IX, Lemmas 3.2, 3.3, and 3.4]).

**Lemma 1.4.** *Let $V, W$ be $p$-permutation $RH$-modules and $K \leq H \leq G$.*

1. *$V{\downarrow}_K^H$ and $V{\uparrow}_H^G$ are $p$-permutation.*
2. *$V \oplus W$ and $V \otimes W$ are $p$-permutation.*
3. *If $V$ is permutation, then $V \cong V^*$ and any direct summand of $V$ is $p$-permutation.*

Since the regular $RG$-module $RG$ is permutation, every projective module is $p$-permutation. Moreover, permutation modules are direct sums of transitive permutation modules, and so there are finitely many isomorphism classes of indecomposable permutation $RG$-modules.

Indecomposable $p$-permutation $RG$-modules are characterised in [23, (0.4)].

**Lemma 1.5.** *Let $M$ be an indecomposable $RG$-module. Then $M$ is $p$-permutation if and only if $M$ is a trivial source module.*

In other words, an indecomposable $p$-permutation module is isomorphic to a direct summand of an induced module $R\uparrow_Q^G$ for some $p$-subgroup $Q$ of $G$, or more generally $R\uparrow_H^G$, where $Q$ is conjugate to a Sylow $p$-subgroup of $H$. By the Alperin–Scott theorem, for each $p$-subgroup $Q$ of $G$, the induced module $k\uparrow_Q^G \cong k[G/Q]$ contains a unique indecomposable direct summand $S_Q(G)$ which has the trivial $kG$-module as socle and as head, and furthermore $S_Q(G) \cong S_Q(G)^*$. We call this summand $S_Q(G)$ the *Scott kG-module associated to* $Q$ (cf. [71]).

Let us assume that $G$ is a finite $p$-group. The notion of a relative syzygy introduced above generalises to $G$-sets instead of subgroups.

**Definition 1.10.** Let $G$ be a finite $p$-group and $X$ a $G$-set. The kernel of the augmentation map $RX \to R$ mapping every element of $X$ to 1 is the relative syzygy $\Omega_X(R)$.

In particular, if $X = G/H$ for some subgroup $H$ of $G$, then $\Omega_{G/H}(R)$ is the $H$-relative syzygy of the trivial module $R$, introduced in Definition 1.7.

Note that in general, the module $\Omega_X(R)$ is not indecomposable. We shall come back to these modules in Sect. 2.3 and later.

## 1.6 Algebraic Topology

By *algebraic topology*, we refer to the study of topological objects using algebra, and vice versa, the study of (abstract) algebraic objects using topology. Until 2016, such a section would have not found its place in a text on endotrivial modules. However, since then, Grodal has elaborated a method based on homotopy theory, in order to investigate the group of endotrivial modules. Hence, in this section, we succinctly present the concepts on which Grodal's work is based, such as the Borel construction and the notion of a homotopy colimit. We will use them in Chap. 5 when we will present his results. We refer the reader seeking a more detailed background and examples on the subject to the extensive, and excellent, resources available in the literature, such as [11, 25, 55, 56, 114].

Beyond endotrivial modules, we often use concepts and techniques from algebraic topology in the analysis of the $p$-local structure of finite groups. By $p$-*local*, we mean the way the $p$-subgroups of $G$ embed and *fuse* into $G$, and how their normalisers "behave" (see Sect. 1.2). Starting from an algebraist's viewpoint, our first step is to associate to a finite group $G$ some topological object(s) intrinsic to $G$. We can, for instance, consider the *classifying space* $BG$ of $G$, or some $G$-*simplicial complex* constructed using the $p$-local structure of $G$.

A *poset* is a **Partially Ordered SET**. Replacing a relation $\leq$ with an arrow $\to$, a poset can also be regarded as a small category whose objects are the set elements and the morphisms sets $\mathrm{Hom}(x, y)$ are empty unless $x \leq y$ in the poset, in which case there is a single map $x \to y$.

### 1.6.1   G-Spaces

We start with a review of some needed topological notions. A suggested reference is
[55].

**Definition 1.11.** Let $G$ be a topological group. A *G-space* is a topological space $X$
together with a $G$-action

$$G \times X \to X \ , \quad (g, x) \mapsto gx$$

subject to $1x = x$ and $(gh)x = g(hx)$ for all $x \in X$ and all $g, h \in G$.

The *orbit space* of such an action is the set of equivalence classes $X/G$ with
respect to the relation: $x \sim y$ *if and only if there exists a* $g \in G$ *such that* $gx = y$,
equipped with the quotient topology. The equivalence classes are the *G-orbits*.

In particular, given $H \leq G$, the natural projection $G \twoheadrightarrow G/H$ allows us to regard
the set of cosets $G/H$ as a topological space with the quotient topology. The $G$-action
on $G/H$ given by left translation, $(g, xH) \mapsto g(xH) = (gx)H$, for all $g, x \in G$,
makes $G/H$ into a $G$-space, and each coset into a $G$-orbit.

A $G$-space $X$ is *transitive* if $X$ is a single $G$-orbit in which case, $X$ is a $G$-space
isomorphic to $G/G_x$, where $G_x$ is the *isotropy group* of $x \in X$, that is, $G_x$ is the
subgroup

$$G_x = \{g \in G \mid gx = x\} \ \text{ of } G.$$

Note that $G_{gx} = {}^g G_x$. We call the $G$-action *free* if $G_x$ is trivial for all $x \in X$.

Given a subroup $H$ of $G$, the *H-fixed points* of $X$ is the subset

$$X^H = \{x \in X \mid hx = x \ , \quad \forall h \in H\} \ \text{ of } X.$$

A subspace $Y$ of $X$ is *G-invariant* if $gy \in Y$ for all $g \in G$ and all $y \in Y$. If $Y$ is
$G$-invariant, then we have an induced action of $G$ on $Y$ turning $Y$ into a $G$-space. A
*G-map* between two $G$-spaces $X$ and $Y$ is a continuous map of topological spaces
$f : X \to Y$ such that $f(gx) = gf(x)$ for all $x \in X$ and all $g \in G$.

If $f : X \to Y$ is a $G$-map, then $f$ induces a $G$-map on the orbit spaces $\bar{f} : X/G \to
Y/G$, since $f(Gx) = Gf(x)$ is a $G$-orbit of $Y$ for all $x \in X$, and $\bar{f}$ is continuous on
the orbit spaces equipped with the quotient topology.

The following is a useful lemma (cf. [55, Proposition 1.14]).

**Lemma 1.6.** *Let $H, K \leq G$. Consider the $G$-spaces $G/H$ and $G/K$.*

1. *There exists a $G$-map $G/H \to G/K$ if and only if $H \leq_G K$.*
2. *Any $G$-map $G/H \to G/K$ is of the form $\mu_g : xH \mapsto xg^{-1}K$ for some $g \in G$
   such that ${}^g H \leq K$, and conversely, each such $g \in G$ gives a well-defined $G$-map
   $\mu_g$. Furthermore, $\mu_g = \mu_{g'}$ if and only if $g'g^{-1} \in K$.*

*Proof.* Note that a $G$-map $f : G/H \to G/K$ must satisfy $f(gH) = gf(1H) =
gf(hH) = ghf(1H)$ for all $g \in G$ and $h \in H$. So $f$ is determined by $f(1H)$, say,

$f(1H) = gK$. The equality $gK = f(hH) = hf(1H) = hgK = g(^{g^{-1}}h)K$ for all $h \in H$ implies that we must have $^{g^{-1}}H \leq K$. In other words, we have shown that if there is a $G$-map $G/H \to G/K$, then it must have the form $\mu_g : hH \mapsto hg^{-1}K$ for all $h \in H$, where $g \in G$ is such that $^gH \leq K$. Clearly each such $\mu_g \in G$ is a $G$-map.

Now, suppose $\mu_g = \mu_{g'}$. Then $g^{-1}K = g'^{-1}K$, and so $g'g^{-1} \in K$. $\qquad\qquad\square$

Recall that two maps $\alpha, \beta : X \to Y$ of topological spaces are *homotopic* if there exists a continuous map $H : X \times [0, 1] \to Y$ such that $H(0, x) = \alpha(x)$ and $H(1, x) = \beta(x)$ for all $x \in X$. Homotopy is an equivalence relation, which we denote by $\alpha \sim \beta$ (cf. [11, Vol II, Sect. 1.2]).

Loosely, a *CW-complex* is a topological space with "nice" combinatorial properties, and constructed by inductively attaching *n-cells* following a certain procedure. In "CW", the "C" means "closure-finite", and the "W" means "weak topology" (cf. [22, Sect. IV.8] for a rigorous definition).

**Definition 1.12.** Let $(S^n, s)$ be the $n$-sphere with basepoint $s$, and let $X$ be a topological space with base point $x$. The set of homotopy equivalence classes of continuous pointed maps $[S^n, s; X, x]$ (i.e. continuous maps with $s \mapsto x$) form the homotopy groups $\pi_n(X, x)$, for all nonnegative integers $n$, where the group structure is defined by subdividing and stretching the interval $[0, 1]$ ([11, Vol II, Proposition 1.2.3]). If $n = 1$, we call $\pi_1(X, x)$ the *fundamental group* of $X$ with *basepoint $x$*.

Given a finite group $G$, the *classifying space* of $G$ is a pointed connected CW-complex $BG$ such that the fundamental group of $BG$ satisfies $\pi_1(BG) \cong G$, whilst $\pi_n(BG) = 0$ for the higher homotopy groups, $n > 1$.

We write $EG$ for the *universal cover* of $BG$. This is a contractible topological space equipped with a free $G$-action.

Given a $G$-space $X$, the *Borel construction* of the $G$-action on $X$ is the quotient

$$X_{hG} = (X \times EG)/G \,, \quad \text{where } G \text{ acts diagonally on } X \times EG.$$

In particular, if $X$ is a point, then $(X \times EG)/G \cong BG$.

### 1.6.2 G-Categories

Finite groups are compact discrete topological spaces. If $G$ is a finite group, then a $G$-set is a $G$-space equipped with the discrete topology. Focussing on this case, we consider the following categories.

The *orbit category* of a finite group $G$ is the category $\mathscr{O}(G)$ whose objects are the $G$-sets $G/H$ for $H \leq G$, and the morphisms are the $G$-maps. This is the full orbit category of $G$. We can specialise it to a family $\mathscr{C}$ of subgroups of $G$ closed under conjugation (i.e. if $H \in \mathscr{C}$, then $^gH \in \mathscr{C}$ for all $g \in G$) and consider the full subcategory $\mathscr{O}_\mathscr{C}(G)$ with objects the quotients $G/H$ for the subgroups $H$ in $\mathscr{C}$. If we only consider nontrivial subgroups, we write $\mathscr{O}_\mathscr{C}^*(G)$. Since we are interested in the $p$-local structure of $G$, we can in particular take the family $\mathscr{S}_p(G)$ (or $\mathscr{S}_p^*(G)$) of all (nontrivial) $p$-subgroups of $G$.

**Definition 1.13.** Given a prime $p$ and a finite group $G$ of order divisible by $p$, we let $\mathscr{O}_p(G)$ denote the *orbit category* of $G$ with objects the $G$-sets $G/Q$, where $Q$ is a $p$-subgroup of $G$ and the morphisms are the $G$-maps. We write $\mathscr{O}_p^*(G)$ for the full subcategory with objects the $G$-sets $G/Q$, where $Q$ is a nontrivial $p$-subgroup of $G$.

In particular, $\mathrm{Hom}_{\mathscr{O}_p(G)}(G/Q, G/R) \neq \emptyset$ if and only if $Q \leq_G R$ and $G/Q \cong G/R$ in $\mathscr{O}_p(G)$ if and only if $Q =_G R$.

Related to the orbit category is the *transport category*, which we can specialise to a family of subgroups of $G$ closed under conjugation. We follow [70, Sect. 2.2].

**Definition 1.14.** The *transport category* is the category $\mathscr{T}(G)$ with objects the subgroups of $G$ and the sets of morphisms $H \longrightarrow K$ between two subgroups $H, K$ of $G$ are the sets

$$\mathrm{Hom}_{\mathscr{T}(G)}(H, K) = \{g \in G \mid {}^g H \leq K\}.$$

We write $\mathscr{T}_p(G)$ for the full subcategory whose objects are the $p$-subgroups of $G$ and $\mathscr{T}_p^*(G)$ if we only consider nontrivial $p$-subgroups.

There is a functor $\mathscr{T}(G) \longrightarrow \mathscr{O}(G)$, given by

$$H \longmapsto G/H \quad \text{and} \quad \big(g \in \mathrm{Hom}_{\mathscr{T}(G)}(H, K)\big) \longmapsto \big(1H \mapsto g^{-1}K\big),$$

on the set of objects, respectively morphisms. This functor restricts to a functor $\mathscr{T}_{\mathscr{C}}(G) \longrightarrow \mathscr{O}_{\mathscr{C}}(G)$ between the full subcategories on $\mathscr{C}$, for any family $\mathscr{C}$ of subgroups of $G$ closed under conjugation.

**Definition 1.15.** Let $\mathscr{C} \subseteq \mathscr{S}_p(G)$ be a family of $p$-subgroups of $G$ closed under conjugation. We define the *fusion* and *fusion-orbit categories* $\mathscr{F}_{\mathscr{C}}(G)$ and $\overline{\mathscr{F}}_{\mathscr{C}}(G)$ to be the categories whose objects are the subgroups in $\mathscr{C}$, and the morphisms between two objects $Q, R \in \mathscr{C}$ are the sets

$$\mathrm{Hom}_{\mathscr{F}_{\mathscr{C}}(G)}(Q, R) = \mathrm{Hom}_{\mathscr{T}_{\mathscr{C}}(G)}(Q, R)/C_G(Q)$$

and

$$\mathrm{Hom}_{\overline{\mathscr{F}}_{\mathscr{C}}(G)}(Q, R) = R \backslash \mathrm{Hom}_{\mathscr{T}_{\mathscr{C}}(G)}(Q, R)/C_G(Q).$$

### 1.6.3   Simplicial Objects and Constructions

Simplicial complexes provide a powerful "combinatorial" technique to study topological spaces. In this section, we present the notions and constructions pertaining to simplicial complexes and sets, which will help our analysis of the $p$-local structure of a finite group $G$. We follow [56], and further background can be found in [22, 25, 114] amongst other references.

Our objective is to equip an algebraic structure with a topology (via an "algebra-topology/topology-algebra dictionary"), and then use the combinatorial description of a topological space using abstract simplices.

**Definition 1.16.** Let $\mathbf{n} = \{0, \dots, n\}$ for a nonnegative integer $n$.

The *topological n-simplex* is the subset

$$\{(x_0, \dots, x_n) \in \mathbb{R}^{n+1} \mid \sum_{0 \le i \le n} x_i = 1\} \subset \mathbb{R}^{n+1}.$$

An *abstract simplicial complex* is a pair $K = (V, S)$, where $V$ are the *vertices* of $K$, and $S$ the *simplices* of $K$. That is, $S$ is a collection of nonempty finite subsets of elements of $V$ closed under taking subsets: if $\sigma \in S$ and $\sigma' \subseteq \sigma$, then $\sigma' \in S$.

An *n-simplex* is a simplex $\sigma \in S$ of cardinality $(n+1)$.

The *geometric* or *topological realisation* of $K$ is the subspace $|K|$ of the real euclidean space $\mathbb{R}^{|V|}$ of all the formal linear combinations $\sum_{v \in V} a_v v$ subject to the following conditions:

- $0 \le a_v \le 1$ for all the vertices $v \in V$,
- $\sum_{v \in V} a_v = 1$, and
- the *support* is itself a simplex, where by the support of a linear combination $\sum_{v \in V} a_v v$, we mean the set $\{v \mid a_v \ne 0\}$ of vertices with nonzero coefficient.

In the category of abstract simplicial complexes, a morphism $f : K \to K'$, for $K = (V, S)$ and $K' = (V', S')$, is a function $f : V \to V'$ such that $f(\sigma) \in S'$ for all $\sigma \in S$. Each morphism $f$ induces a continuous map $f : |K| \to |K'|$ by putting $f(\sum_{v \in V} a_v v) = \sum_{v \in V} a_v f(v)$.

The *abstract n-simplex* is the abstract simplicial complex

$$D_n = (\mathbf{n}, \mathscr{P}(\mathbf{n})),$$

where $\mathscr{P}(\mathbf{n})$ denotes the set of all nonempty subsets of $\mathbf{n}$.

An *ordered simplicial complex* is an abstract simplicial complex $K = (V, S)$, where $V$ is a poset whose order relation restricts to a total order on the simplices: $\sigma = (v_0 \le \dots \le v_n)$. A morphism of ordered simplicial complexes is a morphism of abstract simplicial complexes which preserves the order on the sets of vertices (i.e. if $v_1 \le v_2$ in $K$, then $f(v_1) \le f(v_2)$ in $K'$).

The *ordered n-simplex* is the ordered simplicial complex

$$\Delta_n = (\mathbf{n}, \mathscr{P}(\mathbf{n})),$$

where the order on $\mathbf{n}$ is the standard one.

Note that the geometric realisation of the abstract $n$-simplex is the topological $n$-simplex.

**Definition 1.17.** Let $\Delta$ be the full subcategory of the category of ordered simplicial complexes with objects all the ordered simplices $\Delta_n$, with $n \geq 0$. The morphisms in $\Delta$ are induced by the order preserving maps $\mathbf{n} \to \mathbf{m}$. For all $n \geq 0$ and all $0 \leq i \leq n$, we have the *face* and *degeneracy maps*

$$d_i : \mathbf{n} - 1 \longrightarrow \mathbf{n} \qquad \text{and} \quad s_i : \mathbf{n} + 1 \longrightarrow \mathbf{n}$$
$$j \longmapsto \begin{cases} j & \text{for } j < i \\ j+1 & \text{for } j \geq i \end{cases} \qquad\qquad j \longmapsto \begin{cases} j & \text{for } j \leq i \\ j-1 & \text{for } j > i. \end{cases}$$

In other words, $d_i$ deletes the vertex $i$ and shifts by $-1$ the greater values, while $s_i$ glues the vertices $i$ and $i+1$.

Write $\Delta^{op}$ for the opposite category, that is, the category with the same objects but the arrows reversed.

A *simplicial set* is a covariant functor

$$X : \Delta^{op} \to \mathbf{Sets}.$$

Write **Ss** for the category of simplicial sets and $X_n$ for the set $X(\Delta_n)$ of *n-simplices* of $X$. A morphism in **Ss** is a natural transformation between functors.

More generally, a *simplicial object* in a category $\mathscr{C}$ is a covariant functor $X : \Delta^{op} \to \mathscr{C}$. So we can regard a simplicial object in $\mathscr{C}$ as being a set $\{X_n \mid n \geq 0\}$, where $X_n = X(\Delta_n)$. Write $\Delta X$ for the category of simplices of $X$, where the objects are the simplices and the morphisms are the compositions of face and degeneracy maps.

In [70], the study of endotrivial modules uses the notion of a *homotopy colimit*. To define it, we now need to recap a few notions from homotopy theory, which, for our limited purposes, we restrict to the case of *spaces* $X$ that are simplicial $G$-sets, where $G$ is a finite group. Then, the action of $G$ on the sets $X_n$ of $n$-simplices of $X$ commutes with the face and degeneracy maps.

**Definition 1.18.** *([70, Sect. 2.3])* The *geometric realisation* of a space $X$ is the space

$$|X| = \bigsqcup_{n \geq 0} X_n \times \Delta_n \quad \text{with appropriate identifications (cf. [56, (3.16)]).}$$

The *transport category* of $\Delta X$ is the category $(\Delta X)_G$, whose objects are the simplices of $X$ and the morphisms between two simplices $\sigma, \tau$ in $\Delta X$ are pairs

$$(g, f) \in G \times \mathrm{Mor}(\Delta X) \,, \text{where } f : (g\sigma) \longrightarrow \tau.$$

A *G-local coefficient system* on $X$ is a covariant functor

$$F : (\Delta X)_G \longrightarrow \mathbf{Vect}_k.$$

A *G-twisted coefficient system* is a $G$-local coefficient sending all morphisms of the domain $(\Delta X)_G$ to isomorphisms in $\mathbf{Vect}_k$.

We define the chain complex $C(X; F)$ with $n$-th term $C_n(X; F) = \bigoplus_{\sigma \in \Delta X} F(\sigma)$ and differentials induced by the face maps. The chain complex $C(X; F)$ is a chain complex of $kG$-modules via the $G$-action $F((g, \mathrm{id}_{g\sigma})) : F(\sigma) \longrightarrow F(g\sigma)$.

We can now introduce the notion of a *homotopy colimit*. This construction generalises that of a colimit. In what follows, for our context, it could be useful to think of the $G$-space $X$ as the *nerve* of the small category $\mathscr{C}$. This is the simplicial set whose $n$-simplices are the chains of $n$ composable morphisms in $\mathscr{C}$. If $\mathscr{C}$ is a poset, then an arrow $x \to y$ in $\mathscr{C}$ indicates the relation $x \leq y$. In particular, a 0-simplex is an object of $\mathscr{C}$, a 1-simplex is a morphism in $\mathscr{C}$, and an $n$-simplex is a chain $x_0 < \cdots < x_n$ of $(n+1)$ objects of $\mathscr{C}$. The sets $C_n(\mathscr{C})$ of $n$-simplices, for $n \geq 0$, are $kG$-modules, with respect to the induced action of $G$, and so we obtain a complex $(C_*(\mathscr{C})) = (C_*(\mathscr{C}), d_*)$ of $kG$-modules, where the differential is the alternating sum

$$d_n(x_0 < \cdots < x_n) = \sum_{0 \leq j \leq n} (-1)^j (x_0 < \cdots < \widehat{x_j} < \cdots < x_n) \in C_{n-1}(\mathscr{C}).$$

The *classifying space* of $\mathscr{C}$ is the *topological realisation* of its nerve $|\mathscr{C}|$ (cf. [11, Vol II, Sect. 1.8]). Accordingly, the homotopy groups and, in particular, the *fundamental group* $\pi_1(\mathscr{C})$ of $\mathscr{C}$ are those of $|\mathscr{C}|$.

The homology groups $H_n(\mathscr{C}; k)$ are the homology groups $H_n(C_*(\mathscr{C}); k)$ and the cohomology groups are the cohomology groups of the cochain complex:

$$H^n(C_*(\mathscr{C}); k) = H^n(\mathrm{Hom}_k(C_*(\mathscr{C}), k); k).$$

We can take coefficients in any module, and in particular, $H^1(\mathscr{C}; k^\times)$ is the representation group formed by the isomorphism classes of functors $\mathscr{C} \longrightarrow k^\times$.

If $\sigma = (x_0 < \cdots < x_n)$ is an $n$-simplex, the subgroup $N_G(\sigma) = \bigcap_{0 \leq i \leq n} N_G(x_i)$ is its *normaliser*.

Let $\mathscr{C}$ be a small category and $F : \mathscr{C} \to \mathbf{Ss}$ a simplicial set. Define the simplicial space

$$\bigsqcup_* F \quad \text{where the piece in dimension } n \text{ is} \quad (\bigsqcup_* F)_n = \bigsqcup_{\sigma \in |\mathscr{C}|_n} F(\sigma(0)),$$

where $\sigma = \left( \sigma(0) \xrightarrow{\alpha_1} \sigma(1) \xrightarrow{\alpha_2} \cdots \xrightarrow{\alpha_n} \sigma(n) \right)$.

**Definition 1.19.** Let $\mathscr{C}$ be a small category and $F : \mathscr{C} \to \mathbf{Ss}$ a simplicial set. The *homotopy colimit* of $F$ is the space

$$\mathrm{hocolim}(F) = \mathrm{diag}(\bigsqcup_* F),$$

where diag is the *diagonal functor*.

Explicitly, for each 0-simplex $x_0$ in $|\mathscr{C}|$, we take a copy of $F(x_0)$, for each 1-simplex $x_0 \to x_1$ in $\mathscr{C}$, we take a copy of $F(x_0) \times \Delta[1]$, and for each $n$-simplex $x_0 \to \cdots \to x_n$ in $|\mathscr{C}|$, we take a copy of $F(x_0) \times \Delta[n]$. Then, we make the following identifications ([56, Definition 4.13]):

- if $x_0 \to \cdots \to x_n$ is an $n$-simplex containing an identity map, then we collapse $F(x_0) \times \Delta[n]$;
- we identify the subspace $F(x_0) \times \partial\Delta[n]$ of $F(x_0) \times \Delta[n]$ with the appropriate subspace of $\mathrm{hocolim}(F)$ arising from shorter simplices, where $\partial\Delta[n]$ is the subcomplex of $\Delta[n]$ containing all the simplices except $\{0, \ldots, n\}$.

Given a finite group $G$ and a collection $\mathscr{S}$ of $p$-subgroups of $G$ that is closed under $G$-conjugation, we say that $\mathscr{S}$ is *ample* if the natural map $(|\mathscr{S}|)_{\mathrm{h}\,G} \to BG$ induces an isomorphism in mod-$p$ homology. Here $|\mathscr{S}|$ denotes the nerve of the category associated to $\mathscr{S}$, whose objects are the subgroups in $\mathscr{S}$, and where there is exactly one morphism $x \to y$ for each inclusion $x \leq y$ of subgroups in $\mathscr{S}$ (cf. [56, Example 4.12]).

A *homology decomposition* for $BG$ is an isomorphism in mod-$p$ homology $\mathrm{hocolim}\, F \to BG$ where $F$ is a simplicial set (i.e. a functor from a small category $\mathscr{C}$ to **Sp**) such that for each object $x$ of $\mathscr{C}$, there exists a subgroup $H_x$ of $G$ such that $F(x)$ is weakly equivalent to $BH_x$. Recall that a (co-)chain map is a *weak equivalence* if it induces an isomorphism in mod-$p$ (co-)homology.

There are three well-known homology decompositions: *subgroup*, *centraliser* and *normaliser* decompositions. The last two are those we need for our purposes and they are respectively defined by the functors

$$Q \mapsto C_G(Q) \quad \text{and} \quad Q \mapsto N_G(Q) \quad \text{for each } p\text{--subgroup } Q \text{ of } G.$$

We postpone to Sect. 5.3 the application of these concepts from algebraic topology.

# Chapter 2
# Endotrivial Modules

In this chapter, we introduce the endotrivial modules and the group of endotrivial modules of a finite group, and we also present a few associated notions and elementary facts. There are several excellent surveys on these topics already present in the literature, for instance [30, 31, 120].

## 2.1 Endotrivial Modules

Let $G$ be a finite group and $M$ an $RG$-module. We write $\operatorname{End}_R M = \operatorname{Hom}_R(M, M)$ for the $R$-algebra of $R$-linear transformations $M \to M$. So $\operatorname{End}_R M$ is a (unital) ring with respect to the pointwise addition of maps, $(\varphi + \psi)(m) = \varphi(m) + \psi(m)$ and multiplication given by the composition of maps, $(\varphi\psi)(m) = \varphi(\psi(m))$, for $m \in M$ and $\varphi, \psi \in \operatorname{End}_R M$. By Lemma 1.2, $\operatorname{End}_R M$ is isomorphic to $M^* \otimes M$ as $RG$-modules, where $M^* = \operatorname{Hom}_R(M, R)$ and $\otimes = \otimes_R$. The $G$-action on $M^*$ is $(g\mu)(m) = \mu(g^{-1}m)$, and $G$ acts on $M^* \otimes M$ diagonally, i.e. $g(\mu \otimes m) = g\mu \otimes gm$, for all $g \in G$, $m \in M$ and $\mu \in M^*$.

**Definition 2.1.** An $RG$-module is *endotrivial* if $\operatorname{End}_R M \cong R \oplus (\text{proj})$ in $\operatorname{mod}(RG)$ for some projective $RG$-module denoted (proj). Equivalently, $M$ is endotrivial if $M^* \otimes M \cong R$ in $\operatorname{stmod}(RG)$.

This second approach says that endotrivial modules are the invertible elements in the *Green ring* of the stable module category $\operatorname{stmod}(RG)$. This is the commutative ring whose elements are the stable isomorphism classes of $RG$-modules, with addition given by the direct sum $[M] + [N] = [M \oplus N]$, and multiplication given by the tensor product $[M][N] = [M \otimes N]$ of $RG$-modules. So, an $RG$-module $M$ is invertible if there exists an $RG$-module $N$ such that $M \otimes N \cong R$ in $\operatorname{stmod}(RG)$. In such case, the trivial module $R$ has multiplicity exactly one as a direct summand of

© The Author(s), under exclusive license to Springer Nature Switzerland AG 2019
N. Mazza, *Endotrivial Modules*, SpringerBriefs in Mathematics,
https://doi.org/10.1007/978-3-030-18156-7_2

$M^* \otimes M$, and, by [12, Theorem 2.1], we know that $R|M \otimes N$ if and only if $M \cong N^*$ and $M$ has $R$-rank coprime to $p$.

Given an $RG$-module $M$, there is a surjective evaluation (or trace) map

$$M^* \otimes M \longrightarrow R \ , \quad \mu \otimes m \mapsto \mu(m), \quad \text{for all} \quad \mu \in M^* \text{ and all } m \in M.$$

This map splits if and only if $p$ does not divide the $R$-rank of $M$ (recall that we assume that all our $RG$-modules are $R$-free). So $M$ is endotrivial if and only if the kernel of the evaluation map is projective, and so of dimension divisible by $p$.

**Lemma 2.1.** *Let $M$ be an endotrivial $RG$-module, and let $|G|_p$ denote the $p$-part of $|G|$. Then*

$$\text{rank}_R M \equiv \begin{cases} 1 & (\text{mod } |G|_p) & \text{if } p > 2, \\ 1 & (\text{mod } \frac{1}{2}|G|_2) & \text{if } p = 2. \end{cases}$$

*Example 2.1.* 1. The trivial module $R$ is endotrivial since $R^* \otimes R \cong R$.
2. Let $\Omega_G(R) = \ker(P \twoheadrightarrow R)$ be the kernel of a projective presentation of $R$. Then $\Omega_G(R)$ is endotrivial. Indeed, consider the exact sequence

$$0 \longrightarrow \Omega_G(R) \longrightarrow P \longrightarrow R \longrightarrow 0 \ .$$

Applying the exact functor $\Omega_G(R)^* \otimes$—gives the exact sequence

$$0 \longrightarrow \Omega_G(R)^* \otimes \Omega_G(R) \longrightarrow \Omega_G(k)^* \otimes P \longrightarrow \Omega_G(R)^* \otimes R \longrightarrow 0 \ ,$$

while dualising, i.e. applying the exact contravariant functor $(-)^* = \text{Hom}_R(-, R)$, gives the exact sequence

$$0 \longrightarrow R^* \longrightarrow P^* \longrightarrow \Omega_G(R)^* \longrightarrow 0 \ .$$

Since $\Omega_G(R)^* \otimes P$ and $P^*$ are projective, and since $R^* \cong R$, we obtain by Schanuel's Lemma 1.3 an isomorphism

$$P^* \oplus \left(\Omega_G(R)^* \otimes \Omega_G(R)\right) \cong R \oplus \left(\Omega_G(R)^* \otimes P\right).$$

Since $p \mid |G|$, every projective $RG$-module has dimension divisible by $p$, and so $\text{rank}_R\left(\Omega_G(R)^* \otimes \Omega_G(R)\right) \equiv 1 \pmod{p}$. Therefore $\Omega_G(R)^* \otimes \Omega_G(R)$ is isomorphic to the trivial $RG$-module $R$ in $\text{stmod}(RG)$, by the Krull–Schmidt Theorem 1.2.
Example 2.1 (2) is "THE EXAMPLE" of a nontrivial endotrivial module.

**Proposition 2.1.** *Let $M$, $N$ be $RG$-modules. The following hold.*

1. *Suppose that there exists a projective module $P$ such that $M$ and $N$ fit in an exact sequence $0 \longrightarrow M \longrightarrow P \longrightarrow N \longrightarrow 0$ . Then $M$ is endotrivial if and*

only if $N$ is endotrivial. In particular, $M$ is endotrivial if and only if $\Omega_G^n(M)$ is endotrivial for all $n \in \mathbb{Z}$.

2. Suppose that $M$ is endotrivial. Then $M \cong M_0 \oplus (\text{proj})$ for some indecomposable endotrivial $RG$-module $M_0$.

3. Suppose that $M$ and $N$ are both endotrivial. Then $M \otimes N$ is endotrivial.

The indecomposable endotrivial summand $M_0$ of $M$ in Proposition 2.1 (2) is called the *cap* of $M$.

*Proof.*  1. Apply $M^* \otimes -$ to the exact sequence and compare with the sequence obtained by applying $\operatorname{Hom}_R(-, R) \otimes N$. Both right-hand terms are isomorphic to $M^* \otimes N$ and both middle terms are projective. Hence, Schanuel's lemma yields $M^* \otimes M \cong N^* \otimes N$ in stmod$(RG)$.

2. The assertion is trivially true if $M = M_0$ is indecomposable endotrivial. Suppose that $M = U \oplus V$, then

$$\operatorname{End}_R M \cong \operatorname{End}_R U \oplus \operatorname{End}_R V \oplus \operatorname{Hom}_R(U, V) \oplus \operatorname{Hom}_R(V, U).$$

Note that $R$ is a direct summand of $\operatorname{Hom}_R(U, V)$ if and only if $R$ is a direct summand of $\operatorname{Hom}_R(V, U)$, because $\operatorname{Hom}_R(U, V) \cong U^* \otimes V$ is the $R$-dual of $V^* \otimes U \cong \operatorname{Hom}_R(V, U)$, and so $R$ is a direct summand in either both or none of them. By assumption, $\operatorname{End}_R M \cong R \oplus (\text{proj})$ has a unique trivial direct summand. It follows that exactly one of either $U$ or $V$ is endotrivial and the other projective. Successively eliminating the projective summands, since $M$ has finite rank, we end up with a unique indecomposable endotrivial direct summand of $M$ and the rest is projective.

3. The assertion follows from the isomorphism $(M \otimes N)^* \otimes (M \otimes N) \cong (M^* \otimes M) \otimes (N^* \otimes N)$.

$\square$

Given an $RG$-module $M$, what is an efficient way to find out whether $M$ is endotrivial? Computing tensor products or endomorphism algebras is not a reasonable method, especially when dealing with "large" modules. Obviously, we can first test for the dimension, using Lemma 2.1, and then, if needed, we can apply the following detection result "à la Chouinard" ([42, Theorem 2.9] and [34, Proposition 2.6]).

**Theorem 2.1.** *Let $G$ be a finite group and $M$ an $RG$-module. Then $M$ is endotrivial if and only if $M{\downarrow}_E^G$ is an endotrivial $RE$-module for every elementary abelian $p$-subgroup $E$ of $G$.*

Note that if $M$ is endotrivial, then $M{\downarrow}_H^G$ is endotrivial for any subgroup $H$ of $G$. By contrast, given an endotrivial $RH$-module $V$, the induced module $V{\uparrow}_H^G$ is not endotrivial in general. We shall come back to this fact later.

## 2.2   The Group of Endotrivial Modules

**Definition 2.2.** The set

$$T_R(G) = \{[M] \in \text{stmod}(RG) \mid M^* \otimes M \cong R \text{ in stmod}(RG)\}$$

of stable isomorphism classes of endotrivial $RG$-modules is the *group of endotrivial $RG$-modules*. If there is no confusion about the ring of coefficients $R$, we will write $T(G)$ instead of $T_R(G)$. The group law on $T_R(G)$ is induced by the tensor product of modules,

$$[M] + [N] = [M \otimes N], \quad \text{so that} \quad [R] = 0 \quad \text{and} \quad -[M] = [M^*] \text{ in } T_R(G).$$

*Example 2.2.* In $T_R(G)$, we have $n[\Omega_G(R)] = [\Omega_G^n(R)]$ for all $n \in \mathbb{Z}$. This is clear if $n = 0$, since $\Omega_G^0(R) = R$. If $n > 0$, let $(P_*, \partial_*)$ be a projective resolution of $R$. Then so is the total complex of the tensor product $(P_* \otimes P_*, \partial_* \otimes \partial_*)$. So both are homotopy equivalent, and it follows that

$$\Omega_G^2(R) = \ker(\partial_1) \cong \ker((\partial_* \otimes \partial_*)_1) \cong \Omega_G^1(R) \otimes \Omega_G^1(R) \text{ in stmod}(RG).$$

Iteratively, we obtain the stable isomorphism $\Omega_G^n(R) \cong \Omega_G(R)^{\otimes n}$ for larger $n$. If $n < 0$, then we use the fact that $\Omega_G^n(R) \cong \Omega_G^{-n}(R)^*$, since the dual complex $(P_*^*, \partial_*^*)$ is an injective resolution of $R \cong R^*$.

The group $T_R(G)$ is abelian, and by Proposition 2.1, each stable isomorphism class $[M] \in T_R(G)$ contains a unique indecomposable endotrivial $RG$-module $M_0$. Therefore, when studying $T_R(G)$, it is enough to only consider the indecomposable endotrivial modules.

## 2.3   Endo-Permutation Modules

The word *endotrivial* was coined by Dade in [53, 54], who only considered finite $p$-groups. His motivation resided in studying classes of modules which are potentially classifiable and "useful". By useful, we mean representations which occur often in the representation theory of large classes of groups. In particular, Dade was considering the class of $p$-soluble groups (cf. Sect. 1.2), and he observed in Hall and Higman's article [74] an "interesting" class of modules, which led to the following concept.

**Definition 2.3.** Let $G$ be a finite $p$-group. An $RG$-module $M$ is *endo-permutation* if its endomorphism algebra $\text{End}_R M$ is a permutation $RG$-module. We call $M$ *capped* if $M$ has an indecomposable direct summand $M_0$ with vertex $G$. We call $M_0$ a *cap* of $M$.

The $G$-action is as in Eq. (1.1), i.e. $(g\varphi)(m) = g\varphi(g^{-1}m)$ for $m \in M$, $g \in G$ and $\varphi \in \operatorname{End}_R M$.

In particular, permutation modules are endo-permutation modules, and they are capped if and only if they have a trivial direct summand. Endotrivial modules are capped endo-permutation, since their endomorphism algebra has the form $R \oplus$ (proj) for some projective module (proj). Since $G$ is a finite $p$-group, projective modules are free, and so they are direct sums of copies of $RG$ (which is a permutation $RG$-module).

Intrinsic to endo-permutation modules are their endomorphism algebras. A $G$-algebra over $R$ is an $R$-algebra $A$ such that there exists a group homomorphism $\psi : G \to \operatorname{Aut}(A)$. A prototypical example of a $G$-algebra is the endomorphism algebra of an $RG$-module $M$, where

$$\big(\psi(g)(\varphi)\big)(m) = {}^g\varphi(m) = g\varphi(g^{-1}m), \quad \text{for all} \quad g \in G, \ \varphi \in \operatorname{End}_R M \text{ and } m \in M.$$

We call $A$ *simple* if $A$ is a simple ring, that is, if the only two-sided ideals of $A$ are $0$ and $A$. We say that $A$ is a *permutation algebra* if it has an $R$-basis which makes it into a permutation $RG$-module.

**Definition 2.4.** Let $G$ be a finite $p$-group. A *Dade $G$-algebra* is a simple permutation $G$-algebra $A$ such that

$$\bar{A}(G) = A^G \Big/ \Big( \sum_{H < G} \operatorname{tr}_H^G A + \pi A^G \Big) \neq 0,$$

where $\operatorname{tr}_H^G$ is the relative trace map (Definition 1.6), where $\pi$ is the unique maximal ideal of $R$, and where $A^G$ the set of $G$-fixed points of $A$.

The motivation for introducing this concept is given by one of Dade's observations (but of course, E. Dade did not call such algebras "Dade $G$-algebras", a term which was introduced by Puig in [106]).

**Lemma 2.2.** *Let $G$ be a finite $p$-group and $M$ an $RG$-module. Then $\operatorname{End}_R M$ is a Dade $G$-algebra if and only if $M$ is capped endo-permutation.*

In [53, 54], Dade proves several properties of endo-permutation modules for finite $p$-groups, initiating their classification. In doing so, he introduces the group $\operatorname{Cap}(RG)$ of capped endo-permutation $RG$-modules of a finite $p$-group $G$, which we now call the *Dade group of $G$*. We give here a simplified version of Dade's observations, which summarises the properties we need in our study of endotrivial modules.

**Proposition 2.2.** *Let $G$ be a finite $p$-group and $M, N$ capped endo-permutation modules.*

1. $M \otimes N$ and $M^*$ are capped endo-permutation.
2. Any direct summand of $M$ containing an indecomposable direct summand with vertex $G$ is endo-permutation.
3. If $M_0$, $M_1$ are indecomposable direct summands of $M$ with vertex $G$, then $M_0 \cong M_1$, i.e. a capped endo-permutation module has a unique isomorphism type of caps.
4. $M \oplus N$ is capped endo-permutation if and only if $M$ and $N$ have isomorphic caps, or equivalently if and only if $\mathrm{Hom}_k(M, N)$ is a permutation module with a trivial direct summand.

*Proof.* See [53, Theorem 3.8 and Proposition 3.10].

The above properties of capped endo-permutation modules lead to the definition of an equivalence relation on the class of endo-permutation $RG$-modules: if $M$ and $N$ are capped endo-permutation, then

$$M \sim N \iff M \text{ and } N \text{ have isomorphic caps.}$$

**Definition 2.5.** The *Dade group of G (over R)* is the set of equivalence classes of capped endo-permutation $RG$-modules

$$D_R(G) = \{[M] \mid M \text{ is a capped endo-permutation } RG\text{-module}\}$$

with composition law
$$[M] + [N] = [M \otimes N].$$

If there is no confusion about the ring of coefficients $R$, we write $D(G)$ instead of $D_R(G)$.

In particular, $0 = [R]$ is the equivalence class formed by all the permutation $RG$-modules with a trivial direct summand, and $-[M] = [M^*]$ for a capped endo-permutation $RG$-module $M$. Therefore the group $T(G)$ of endotrivial $RG$-modules is isomorphic to the subgroup of $D(G)$ formed by all the equivalence classes whose cap is endotrivial. However, if $M$ is endotrivial, then not every module in $[M]$ is endotrivial. We will see an alternative interpretation of $T(G)$ as a subgroup of $D(G)$ at the end of Sect. 2.3.2.

In [3], J. Alperin describes the "basic model" of an element of the Dade group of a finite $p$-group, and he also obtains several useful results, such as the torsion-free rank of the group of endotrivial modules. Although Alperin states his results for $R = k$, they hold for $R = \mathscr{O}$ too. Recall from Definition 1.10 that if $X$ is a finite $G$-set, then $\Omega_X(R) = \ker(RX \to R)$ is the kernel of the augmentation map sending every element of $X$ to 1. Note that $\Omega_X(\mathscr{O})$ is an $\mathscr{O}G$-module lifting the $kG$-module $\Omega_X(k)$.

**Theorem 2.2.** ([3]) *Let $G$ be a finite $p$-group and $X$ a finite $G$-set. The relative syzygy $\Omega_X(R)$ is endo-permutation. Furthermore,*

1. $\Omega_X(R)$ is capped if and only if $|X^G| \neq 1$.
2. $\Omega_X(R)$ is indecomposable if and only if no orbit of $G$ on $X$ is a homomorphic image, as a $G$-set, of another orbit of $G$ on $X$.
3. If $\Omega_X(R)$ is indecomposable and if $H$ is a normal subgroup of $G$, then $\Omega_{X^H}(R)$ is a capped endo-permutation $R[G/H]$-module.

Here is a key result obtained by Puig in [107, Corollary 2.4], and to which we will return in Sect. 3.1.

**Theorem 2.3.** *Let $G$ be a finite $p$-group. Then, the Dade group $D(G)$ of $G$ is finitely generated.*

In other words, $D(G) \cong DT(G) \oplus DF(G)$, where $DT(G)$ is a finite abelian group and $DF(G)$ is a torsionfree abelian group of finite rank.

The successive articles [18, 21] achieve the classification of endo-permutation modules for any finite $p$-group, by adopting a functorial approach.

## 2.3.1 Generalisation of Endo-Permutation Modules

Attempts to generalise the Dade group for an arbitrary finite group have not gone very far. The most natural one, by Urfer in [121], introduced the concept of an *endo-p-permutation module*, mimicking that of a $p$-permutation module. That is, given an arbitrary finite group $G$, an $RG$-module is endo-$p$-permutation if its restriction to every $p$-subgroup of $G$ is endo-permutation. Urfer obtained a result characterising the indecomposable endo-$p$-permutation modules with vertex a Sylow $p$-subgroup $S$ of $G$ via their source modules, and he also classified the sources of endo-$p$-permutation modules when the normaliser of $S$ controls $p$-fusion using the classification of endo-permutation modules for a finite $p$-group.

In [94], the authors elaborate on the categorical approach by introducing the *Dade group of a fusion system*.

A *(saturated) fusion system* on a finite $p$-group $S$ is the category whose objects are the subgroups of $S$ and the morphisms are injective group homomorphisms subject to the five axioms below, encoding the way in which a finite group acts by conjugation on its $p$-subgroups:

1. composition of morphisms in $\mathscr{F}$ is the usual composition of group homomorphisms;
2. if $\varphi : Q \to Q'$ is a morphism in $\mathscr{F}$ then so is the induced isomorphism $Q \cong \varphi(Q)$ as well as its inverse;
3. $\mathrm{Hom}_{\mathscr{F}}(Q, Q')$ contains the set $\mathrm{Hom}_S(Q, Q')$ of group homomorphisms $\varphi : Q \to Q'$ for which there exists an element $y \in S$ satisfying $\varphi(u) = yuy^{-1}$ for all $u \in Q$;
4. (I-S) if $|N_S(Q)| \geq |N_S(\varphi(Q))|$ for any $\varphi \in \mathrm{Hom}_{\mathscr{F}}(Q, S)$ then $\mathrm{Aut}_S(Q)$ is a Sylow $p$-subgroup of $\mathrm{Aut}_{\mathscr{F}}(Q)$;

5. (II-S) if $\varphi : Q \to S$ is a morphism in $\mathscr{F}$ such that $|N_S(\varphi(Q))| \geq |N_S(\tau(Q))|$ for
   any $\tau \in \mathrm{Hom}_{\mathscr{F}}(Q, S)$ then $\varphi$ extends to a morphism $\psi : N_\varphi \to S$ in $\mathscr{F}$, where
   $N_\varphi$ is the subgroup of $N_S(Q)$ consisting of all $y \in N_S(Q)$ for which there exists
   a $z \in N_S(\varphi(Q))$ with the property $\varphi(yuy^{-1}) = z\varphi(u)z^{-1}$ for all $u \in Q$.

**Definition 2.6.** Let $S$ be a finite $p$-group and let $\mathscr{F}$ be a fusion system on $S$. The
*Dade group* $D(S, \mathscr{F})$ of $(S, \mathscr{F})$ is the abelian group

$$D(S, \mathscr{F}) = \varprojlim_{\mathscr{F}} \mathbf{D} ,$$

where $\mathbf{D}$ denotes the contravariant functor from $\mathscr{F}$ to the category of abelian groups
sending a subgroup $Q$ of $S$ to the Dade group $D(Q)$ and a morphism $\varphi : Q \to P$ in
$\mathscr{F}$ to the group homomorphism $\mathrm{Res}_\varphi : D(P) \to D(Q)$. Define

$$T(S, \mathscr{F}) = \varprojlim_{\mathscr{F}} \mathbf{T} \quad \text{and} \quad DT(S, \mathscr{F}) = \varprojlim_{\mathscr{F}} \mathbf{DT} ,$$

where $\mathbf{T}$ and $\mathbf{DT}$ are the subfunctors of $\mathbf{D}$ sending a subgroup $Q$ of $S$ to $T(Q)$ and
$DT(Q)$, respectively.

The Dade group of the category $\mathscr{F}$ uses *inverse (or projective) limits* of a *coefficient system* (cf. [11, Vol II, Sects. 7.1 and 7.2]). In our case, this coefficient system
is $\mathbf{D}$, which associates an abelian group $D(Q)$ to each object $Q \in \mathscr{F}$, and it is subject
to the condition that for each morphism $\varphi : Q \to P$ in $\mathscr{F}$, we have a commutative
diagram

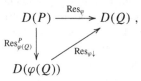

where $\mathrm{Res}^P_{\varphi(Q)}$ is the restriction along the inclusion $\varphi(Q) \leq P$, and where $\varphi \downarrow$ is the
induced isomorphism $Q \to \varphi(Q)$. The inverse limit of such a coefficient system is an
abelian group $D(S, \mathscr{F})$ together with group homomorphisms $\mathrm{Res}_Q : D(S, \mathscr{F}) \to$
$D(Q)$ for each subgroup $Q$ of $S$, which commute with all the restriction maps
along the morphisms in $\mathscr{F}$, and $D(S, \mathscr{F})$ is *universal* in the sense that, given any
abelian group $X$ and maps $r_Q : X \to D(Q)$, one for each $Q \leq S$, that commute
with the restrictions along the morphisms in $\mathscr{F}$, there is a unique group homomorphism $\alpha : X \to D(S, \mathscr{F})$ such that $r_Q = \mathrm{Res}_Q \circ \alpha : X \to D(Q)$. In other words,
$\left(D(S, \mathscr{F}), (\mathrm{Res}_Q)_{Q \leq S}\right)$ is the smallest "system" with the above stipulated property,
and each other "system" $(X, (r_Q)_{Q \leq S})$ must factor through $\left(D(S, \mathscr{F}), (\mathrm{Res}_Q)_{Q \leq S}\right)$.

We identify $D(S, \mathscr{F})$ with a subgroup of $D(S)$. It turns out that the elements in
$D(S, \mathscr{F})$ are the $\mathscr{F}$-stable equivalence classes of endo-permutation $RS$-modules,
and moreover, in each $\mathscr{F}$-stable element in $D(S)$, there exists an $\mathscr{F}$-stable endo-
permutation $RS$-module (cf. [94, Propositions 3.4 and 3.7]).

The Dade group of a fusion system $\mathscr{F}$ on a finite $p$-group $S$ generalises the notion of the Dade group of $S$ because if $\mathscr{F} = \mathscr{F}_S(S)$ is the trivial fusion system defined by the conjugation action of $S$ on its subgroups, then $D(S, \mathscr{F}_S(S)) \cong D(S)$. More generally, if $\mathscr{F} = \mathscr{F}_S(G)$ for $G$ a finite group with $S \in \mathrm{Syl}_p(G)$, then Urfer shows in [121, Proposition 2.19] that the restriction from $G$ to $S$ induces a group isomorphism between the group of endo-$p$-permutation $RG$-modules with vertex $S$ and $D(S, \mathscr{F}_S(G))$. An $RG$-module is *(capped) endo-p-permutation* if its restriction to a Sylow $p$-subgroup $S$ of $G$ is (capped) endo-permutation. The capped endo-$p$-permutation modules can be made into an abelian group by taking the equivalence classes with respect to the relation: $M \sim N$ *if* $M{\downarrow}_S^G$ *and* $N{\downarrow}_S^G$ *have isomorphic caps.*

The two main motivations for studying the Dade group of a fusion system are the *detection* (from global to local) and the *gluing* (from local to global) problems. That is, in our case, they address the questions:

- Detection: *"Is there a "minimal" class of p-groups $\mathscr{X}$ such that given $x \in$ mod $(kS)$ for $S$ a finite p-group, then $x \in D(S)$ provided $\mathrm{Res}_Q^S x \in D(Q)$ for each $Q \in \mathscr{X}$?"*
- Gluing: *"Given a collection of local data $\{x_{Q/Q'} \in D(Q/Q') \mid Q' \trianglelefteq Q \leq S\}$ such that $\mathrm{Defres}_{U/U'}^{Q/Q'}(x_{Q/Q'}) = x_{U/U'}$ whenever $Q' \leq U' \trianglelefteq U \leq Q \leq S$ and $Q' \trianglelefteq Q$, is there an $x \in D(S)$ such that $\mathrm{Defres}_{Q/Q'}^S x = x_{Q/Q'}$ for all $Q/Q'$? If so, can we find a minimal class of p-groups $\mathscr{C}$ such that it suffices to be given a collection of local data $x_{Q/Q'}$ with $Q/Q' \in \mathscr{C}$ to obtain the existence and uniqueness of $x \in D(S)$?"*

(We will define the deflation-restriction map Defres in Theorem 2.8.)

In the case of endo-permutation modules, Bouc, in [19], addresses the gluing question by means of an exact sequence, stating that the gluing problem always has a solution whenever $\mathrm{H}^1(\mathscr{E}_{\geq 2}(S), \mathbb{Z}) = 0$, but we cannot say much otherwise.

In the case of the Dade group of a fusion system, the authors in [94] obtain that $D(S, \mathscr{F})$ is detected upon restriction to the class of all $p$-groups of order at most $p^3$ and exponent at most $p$. On the other hand, denoting by $\mathscr{F}^*$ the full subcategory of $\mathscr{F}$ formed by the nontrivial subgroups of $S$, and $\mathbb{E}_S$ the quotient of the $\mathbb{F}_2$-vector space of maps $\mathscr{E}_{\geq 2}(S) \to \mathbb{F}_2$ that are constant on the connected components of $\mathscr{E}_{\geq 2}(S)$ by the constant maps, there is a short exact sequence (the "gluing sequence")

$$0 \longrightarrow DT(S, \mathscr{F}) \longrightarrow \mathbf{D}(S, \mathscr{F}^*) \longrightarrow \mathbb{E}_S \longrightarrow 0 ,$$

where the middle term is the inverse limit of the Dade groups of the subquotients of $S$ of order at most $p^3$ and exponent at most $p$.

Let us now turn to the generalisation of the Dade group over the field $k$ that Lassueur studied in [86]. The idea is to substitute "projectivity" with "relative projectivity with respect to a $kG$-module" in the definition of endotrivial modules. Recall from Definition 1.8 that an $RG$-module $M$ is $V$-*projective* if there exists a $kG$-module $N$ such that $M \mid V \otimes N$.

**Definition 2.7.** Let $G$ be a finite group and $V, M$ two $kG$-modules. $M$ is $V$-*endotrivial* if $\mathrm{End}_k M \cong k \oplus (\mathrm{proj})_V$ for some $V$-projective module $(\mathrm{proj})_V$.

The trivial module $k$ is $V$-projective if and only if there is at least one indecomposable direct summand of $V$ whose dimension is coprime to $p$, since then $k$ is a direct summand of $V \otimes V^*$. (Recall that [12, Theorem 2.1] asserts that the trivial module $k$ has multiplicity exactly one as a direct summand of a tensor product of $kG$-modules $M \otimes N$ if and only if $M \cong N^*$ and $M$ has dimension coprime to $p$.) If so, then all the $kG$-modules are $V$-projective and so all modules are $V$-endotrivial. Thus, the focus is on *absolutely $p - divisible$* modules $V$, that is, each indecomposable direct summand of $V$ has dimension divisible by $p$. Assuming that $V$ is absolutely $p$-divisible, Lassueur extends many results from endotrivial modules to $V$-endotrivial modules. In particular, the set $T_V(G)$ of $V$-endotrivial modules forms an abelian group with respect to the addition induced by $\otimes$.

To regard $T_V(G)$ as a generalisation of the Dade group $D(G)$ for a finite $p$-group $G$, we start by observing that $M$ is an indecomposable endo-permutation $kG$-module if and only if

$$\mathrm{End}_k M \cong k \oplus \bigoplus_j k\!\uparrow_{H_j}^G \quad \text{for some proper subgroups} \quad H_j \text{ of } G.$$

Thus, putting $V = \oplus_j k\!\uparrow_{H_j}^G$, we see that $M$ is $V$-endotrivial. As a consequence, Lassueur proves the following in [86, Theorem 5.0.2].

**Theorem 2.4.** *Let $G$ be a finite $p$-group. The Dade group $D(G)$ is isomorphic to a subgroup of the group of $V$-endotrivial modules, where $V = \displaystyle\bigoplus_{1 < H < G} k\!\uparrow_H^G$.*

## 2.3.2 Operations on the Dade Group and Functoriality

The elements in the Dade group of a finite $p$-group $G$ are mainly "linear combinations" of equivalence classes of relative syzygies. It is therefore convenient to introduce the following notation.

**Notation 2.5.** *Let $G$ be a finite $p$-group and $X$ a finite $G$-set. We denote by $\Omega_X$ the equivalence class of $\Omega_X(R)$ in $D(G)$. Given two $G$-sets $X$ and $Y$, we write $X \sqcup Y$ for their disjoint union. If $H \leq G$ and $M$ is a $kH$-module, we write $\Omega_H(M)$ for the kernel of a $kH$-projective cover of $M$. Note that if $H = G$, then $G = G/1$ as a $G$-set and $\Omega_G = [\Omega_G(k)]$.*

To study endo-permutation and endotrivial modules, we first compare relative syzygies in $D(G)$ with those of the Dade groups of subquotients of $G$. We also want to find out how addition of syzygies works. In this section we address the questions:

1. We know that every $G$-set $X$ can be written as a disjoint union $X = \sqcup_j G/H_j$ for subgroups $H_j$ of $G$. Can we express $\Omega_X \in D(G)$ in terms of the elements $\Omega_{G/H_j}$? If so, how?
2. If $\Omega_X \in D(H)$ for $H \leq G$, then we know that $\mathrm{Ten}_H^G \Omega_X \in D(G)$, by [21, Sect. 2], but can we express this element as a $\mathbb{Z}$-linear combination of relative syzygies in $D(G)$? If so, how?

Theorems 2.6 and 2.7 answer these questions, providing in this way useful formulae for the group $D(G)$.

**Theorem 2.6.** ([99, Lemme 1.5.8–Lemme de Thévenaz])
*Let $G$ be a finite $p$-group and $X_1, \ldots, X_n$ finite $G$-sets. Put $Y = \displaystyle\bigsqcup_{1 \leq i \leq n} X_i$. Then*

$$\Omega_Y = \sum_{1 \leq s \leq n} (-1)^{s-1} \Big( \sum_{1 \leq i_1 < \cdots < i_s \leq n} \Omega_{Y_{i_1, \ldots, i_s}} \Big) \quad \text{in } D(G),$$

*where*

$$Y_{i_1, \ldots, i_s} = \prod_{1 \leq j \leq s} X_{i_j} \quad \text{for all } 1 \leq i_1 < \cdots < i_s \leq n \quad \text{and} \quad 1 \leq s \leq n.$$

*In particular,*
$$\Omega_{X \sqcup Y} = \Omega_X + \Omega_Y - \Omega_{X \times Y}.$$

We have seen in Sect. 1.3.1 some operations on module categories and their functorial properties. In particular, if $H \leq G$, then we have the restriction $\mathrm{Res}_H^G : \mathrm{mod}\,(RG) \to \mathrm{mod}\,(RH)$, the usual induction $\mathrm{Ind}_H^G : \mathrm{mod}\,(RH) \to \mathrm{mod}\,(RG)$, and the tensor induction $\mathrm{Ten}_H^G : \mathrm{mod}\,(RH) \to \mathrm{mod}\,(RG)$. If $K \trianglelefteq G$, then we have the inflation $\mathrm{Inf}_{G/K}^G : \mathrm{mod}\,(R[G/K]) \to \mathrm{mod}\,(RG)$. Some of these operations and properties induce homomorphisms between corresponding Dade groups of finite $p$-groups or groups of endotrivial modules of finite groups.

**Proposition 2.3.** *1. Let $G$ be a finite $p$-group and $H$, $K$ subgroups of $G$ with $K \trianglelefteq G$. Then the functors $\mathrm{Res}_H^G$, $\mathrm{Ten}_H^G$ and $\mathrm{Inf}_{G/K}^G$ induce group homomorphisms*

$$D(G) \underset{\mathrm{Ten}_H^G}{\overset{\mathrm{Res}_H^G}{\rightleftarrows}} D(H) \quad \text{and} \quad \mathrm{Inf}_{G/K}^G : D(G/K) \to D(G).$$

2. *Let $G$ be a finite group and $H$, $K$ subgroups of $G$ with $K \trianglelefteq G$. Suppose that $p \mid |H|$ and $p \nmid |K|$. Then the functors $\mathrm{Res}_H^G$ and $\mathrm{Inf}_{G/K}^G$ induce group homomorphisms*

$$\mathrm{Res}_H^G : T(G) \longrightarrow T(H) \quad \text{and} \quad \mathrm{Inf}_{G/K}^G : T(G/K) \to T(G).$$

*Moreover, if H is strongly p-embedded in G, then induction induces a group isomorphism* $\mathrm{Ind}_H^G : T(H) \to T(G)$.

Induction does not induce a group homomorphism $D(H) \to D(G)$ (resp. $T(H) \to T(G)$) for $H < G$. Indeed, $\mathrm{Ind}_H^G k = k[G/H]$ is not capped endo-permutation (resp. endotrivial) as a $kG$-module.

The next result can be seen as a "change of basis" in $D(G)$.

**Theorem 2.7.** ([17, Theorem 5.2–Formule de Bouc])
*Let H be a subgroup of a finite p-group G and X a non-empty finite H-set. Then,*

$$\mathrm{Ten}_H^G \, \Omega_X = \sum_{U \leq_G V} \mu_G(U, V) |\{a \in V \backslash G/H \mid X^{V^a \cap H} \neq \emptyset\}| \Omega_{G/U} \quad in \, D(G),$$

*where U and V run through a set of representatives of the G-conjugacy classes of subgroups of G, and* $\mu_G(U, V)$ *is the Möbius function defined inductively on the poset of conjugacy classes of subgroups of the finite p-group G by*

$$\mu_G(U, U) = 1 \, , \; \mu_G(U, V) = 0 \;\; if \;\; U \not\leq V, \, and \, \mu_G(U, V) = \sum_{U \leq_G T <_G V} \mu_G(U, T),$$

*for all subgroups U, V of G.*

A more subtle operation on the Dade group is that of *deflation*, which Dade called the *slash operation*. To define it, we need to work with the $k$-endomorphism algebra $\mathrm{End}_k M$ of an endo-permutation $kG$-module $M$ (and so $R = k$).

**Definition 2.8.** Let $H$ be a normal subgroup of $G$ and $A$ a $G$-algebra over $k$. The *deflation* from $G$ to $G/H$ of $A$ is the $k[G/H]$-algebra

$$\mathrm{Def}_{G/H}^G A = A^H / \sum_{K < H} \mathrm{tr}_K^H(A).$$

Recall from Definition 1.6 that the relative trace map $\mathrm{tr}_K^H : A^K \to A^H$ takes $a \in A^K$ to $\mathrm{tr}_K^H(a) = \sum_{g \in [H/K]} {}^g a \in A^H$.

If $G$ is a finite $p$-group and we apply the deflation map $\mathrm{Def}_{G/H}^G$ to $A = \mathrm{End}_k M$ with $M$ capped endo-permutation, it turns out that $\mathrm{Def}_{G/H}^G A \cong \mathrm{End}_k V$, where $V$ is a capped endo-permutation $k[G/H]$-module ([53, Sects. 4 and 5]). Moreover, $V$ is uniquely defined up to equivalence in $D_k(G/H)$. This "uniqueness" assertion is the reason why we work over $k$ instead of $\mathcal{O}$. To palliate this drawback, it suffices to observe that the kernel of the group homomorphism $D_{\mathcal{O}}(G/H) \to D_k(G/H)$ induced by reduction modulo the unique maximal ideal $\pi$ of $\mathcal{O}$ is formed by the $\mathcal{O}[G/H]$-modules of $\mathcal{O}$-rank one.

Let us take a closer look at the functorial properties of relative syzygies explored in [17, Sect. 4].

**Proposition 2.4.** *Let* $G, H$ *be finite* $p$*-groups and* $X$ *a finite* $G$*-set. If* $\varphi \in$ $\mathrm{Hom}(H, G)$*, then*

$$\mathrm{Res}_\varphi(\Omega_X) = (\Omega_{\mathrm{Res}_\varphi X}) \ \text{ in } D(H).$$

*In particular, if* $H, K \leq G$ *and* $K \trianglelefteq G$*, then*

$$\mathrm{Res}^G_H \Omega_X = \Omega_{X \downarrow^G_H} \ \text{ in } D(H), \text{ and } \ \mathrm{Inf}^G_{G/K} \Omega_Y = \Omega_Y \in D(G),$$

*for a* $G/K$*-set* $Y$ *which we can regard as a* $G$*-set with trivial* $K$*-action.*

Let us emphasise that these equations hold in some Dade group, not as $RG$-modules, i.e. up to isomorphism and direct sums of nontrivial indecomposable permutation modules.

As we have seen above in Bouc's formula, Theorem 2.7, tensor induction of relative syzygies yields a more complicated form. By contrast, the deflation operation on relative syzygies is easier ([17, (4.6) Lemma]).

**Lemma 2.3.** *Let* $G$ *be a finite* $p$*-group,* $K \trianglelefteq G$ *and* $X$ *a finite* $G$*-set. Then the deflation from* $G$ *to* $G/K$ *induces a group homomorphism* $\mathrm{Def}^G_{G/K} : D(G) \to D(G/K)$*. Moreover,*

$$\mathrm{Def}^G_{G/K} \Omega_X = \Omega_{X^K} \ \text{ in } \ D(G/K),$$

*with the convention that* $\Omega_\emptyset = 0$ *in any Dade group.*

In particular, $\mathrm{Def}^G_{G/K} \Omega_G = 0 \in D(G/K)$. This leads to the conclusion of this subsection with another characterisation of endotrivial modules.

**Theorem 2.8.** ([107, Eq. (2.1.1)])
*Let* $G$ *be a finite* $p$*-group. Then, as a subgroup of* $D(G)$*,*

$$T(G) = \bigcap_{1 < H < G} \ker \left( \mathrm{Defres}^G_{N_G(H)/H} : D(G) \longrightarrow D(N_G(H)/H) \right),$$

*where* $\mathrm{Defres}^G_{N_G(H)/H} = \mathrm{Def}^{N_G(H)}_{N_G(H)/H} \mathrm{Res}^G_{N_G(H)}$.

### 2.3.3 Endo-Permutation as Sources of Simple Modules for p-Soluble Groups

As mentioned at the beginning of Sect. 2.3, endo-permutation modules were singled out by E. Dade in his study of the modular representations of $p$-soluble groups. The following theorem supports Dade's claim (cf. [118, Theorem 30.5]).

**Theorem 2.9.** *Let* $G$ *be a finite* $p$*-soluble group and* $M$ *a simple* $kG$*-module with vertex* $Q$ *and source* $V$*. Then* $V$ *is an endo-permutation* $kQ$*-module.*

In [61, 2.8], W. Feit conjectured that, for any finite $p$-group $Q$, there are finitely many isomorphism classes of $kQ$-modules $V$ which can be sources of simple modules; in the sense that there exists a finite group $G$ with a $p$-subgroup isomorphic to $Q$, and a simple $kG$-module $M$ with vertex $Q$ and source isomorphic to $V$. The conjecture has been proved in the case when the dimension of the source is bounded ([118, Theorem 30.8]), and Puig proved the conjecture in the case of finite $p$-soluble groups ([107], see also [118, Theorem 30.8]). In fact, in [106, Théorème 8.39 and Remarque 8.41], L. Puig describes the possible endo-permutation modules which can be sources of simple modules for a finite $p$-soluble group. In [100, Theorem], we prove that each module which can be a source of a simple module for a finite $p$-soluble group can be realised as a source of a simple module for a finite $p$-nilpotent group.

**Theorem 2.10.** *Let S be a finite p-group. Let $\mathscr{S}$ be a set of subquotients of S that are cyclic, quaternion, or semi-dihedral. Then any indecomposable endo-permutation kS-module M with vertex S that is a direct summand of a kS-module of the form*
$$\bigotimes_{A/B \in \mathscr{S}} \mathrm{Ten}^S_A \, \mathrm{Inf}^A_{A/B} \, M_{A/B},$$ *where $M_{A/B}$ is an indecomposable torsion endo-permutation module with vertex $A/B$, can explictly be realised as the source of a simple kG-module with vertex S for a finite p-soluble group G.*

Instead of a (sketch of a) proof, let us give an example.

*Example 2.3.* We want to construct a simple $kG$-module $M$ with a cyclic vertex $S$ and source $\Omega_S(k)$ for a $p$-nilpotent group $G$ containing $S$. Let $r$ be a prime different from $p$ and $S$ a cyclic $p$-group such that $r \equiv |S| - 1 \pmod{p|S|}$. Let $Q$ be an extra-special $r$-group of order $r^3$ and exponent $r$. We know that $\mathrm{Aut}(Q/Z(Q)) \cong \mathrm{SL}_2(\mathbb{F}_r)$, of order $r(r^2 - 1)$, and so we can let $S$ act nontrivially on $Q$, and because $r \neq p$, the action is split. Hence we obtain a $p$-nilpotent group $G = Q \rtimes S$. Note that $S$ is a trivial intersection subgroup of $G$.

Write $z$ for a generator of $Z(Q)$, and choose an elementary abelian subgroup $A = \langle x, z \rangle$ of $Q$ of rank 2. Choose $k$ large enough so that $k$ contains a primitive $(r - 1)$st root of unity $\zeta$. Define a representation $\psi : A \to k^\times$ such that $\psi(x) = 1$ and $\psi(z) = \zeta$ on the generators of $Q$. Then $\psi\uparrow^Q_A$ is irreducible of degree $r$, and we call $X$ the corresponding $kQ$-module. By [60, Theorem IX.4.1], there exists a unique $kG$-module $M$ of dimension $r$, up to isomorphism, such that $M\downarrow^G_Q \cong X$, and moreover, $M\downarrow^G_Q$ is endo-permutation. Using Mackey's formula, we see that $\Omega_Q(k)$ is a source for $M$ (cf. [100, Sect. 4]).

Related to this result are [76, 93], in which the authors find endo-permutation modules in their analysis of the local structure of the blocks of finite group algebras. In [76], endo-permutation modules are key in the proof of the main result of the article, namely the proof of the existence of a splendid tilting complex leading to a Morita equivalence inducing a derived equivalence isomorphic to a splendid derived equivalence between a block of a finite $p$-soluble group $G$ with abelian defect group $D$ and its Brauer correspondent for $N_G(D)$. As pointed out by Harris and Linckelmann, the result can be generalised to non $p$-soluble groups and nonabelian defect groups, as the proofs mainly rely on the existence of endo-split $p$-permutation resolutions (and hence an underlying endo-permutation module). In [93], M. Linckelmann proves that, given a bimodule $M$ between two blocks of finite group algebras with a common defect group $D$ and the same fusion system $\mathscr{F}$ on $D$, if $M$ has an endo-permutation source which is $\mathscr{F}$-stable and if $M$ induces a Morita equivalence between the centralisers of nontrivial subgroups of $D$, then $M$ induces a stable equivalence of Morita type. In other words, if $A$ and $B$ are our two blocks and $M$ is an $A - B$-bimodule, then there exists a $B - A$-bimodule $N$ such that $M$ and $N$ are finitely generated projective as one-sided modules and we have $M \otimes_B N \cong A \oplus P$ and $N \otimes_A M \cong B \oplus P'$ for some projective $A$- and $B$-bimodules $P$ and $P'$ respectively.

## 2.4 Endotrivial Modules and the Auslander–Reiten quiver

Auslander and Reiten developed several tools useful in the study of Artin algebras (cf. [6]). The two that are mainly used in our work are those of an *almost split sequence* (also called an Auslander–Reiten, or AR-sequence), and that of the Auslander–Reiten quiver. They are closely related by definition. An idiosyncrasy of endotrivial modules, and more generally endo-$p$-permutation and relative endotrivial modules, is their position in the Auslander–Reiten quiver of the group algebra (cf. [13, 87]). In this section, we state these results, together with a very brief synthesis of the needed theoretical background, referring the reader to [6], or [10, Sects. 2.17 and 2.28]. In the specific case of group algebras, [57, 122] describe the tree class of a connected stable Auslander–Reiten quiver that one can obtain, and it turns out that they can be classified into finite graphs, euclidean diagrams, or infinite trees.

**Definition 2.9.** Let $M$, $N$ be two indecomposable $RG$-modules.

An *almost split sequence* is a nonsplit short exact sequence

$$0 \longrightarrow M \longrightarrow X \xrightarrow{\pi} N \longrightarrow 0$$

such that whenever there exist an $RG$-module $L$ and a nonsplit surjective map $\varphi \in \operatorname{Hom}_{RG}(L, N)$, then $\varphi$ factors through $\pi$, i.e. there exists a $\phi$ such that the diagram

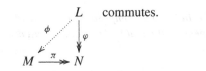

commutes.

A map $\varphi \in \operatorname{Hom}_{RG}(M, N)$ is *irreducible* if $\varphi$ is not an isomorphism, and whenever there is a factorisation $\varphi = \beta\alpha$ of $RG$-homomorphisms, then either $\beta$ has a left inverse or $\alpha$ has a right inverse.

The *Auslander–Reiten quiver* of $G$ is the directed graph whose vertices are the isomorphism classes of indecomposable $RG$-modules, where $R = \mathcal{O}$ or $R = k$, and in which each arrow $M \longrightarrow N$ between two indecomposable $RG$-modules is an irreducible morphism. The edges in the Auslander–Reiten quiver are labeled by pairs of integers as defined in [10, Sect. 2.28].

The *stable Auslander–Reiten quiver* is the modified Auslander–Reiten quiver of $RG$ obtained by deleting the projective modules.

For our purposes, we consider the case when $R = k$, in which case, the ubiquitous almost split sequence takes the form

$$0 \longrightarrow \Omega_G(M) \longrightarrow X \longrightarrow \Omega_G^{-1}(M) \longrightarrow 0 .$$

[5, Proposition 4.11] asserts that whenever the middle term in an almost split sequence contains an indecomposable projective direct summand $P$, then the sequence is of the form

$$0 \longrightarrow \mathrm{Rad}(P) \longrightarrow \big(\mathrm{Rad}(P)/\mathrm{Soc}(P)\big) \oplus P \longrightarrow P/\mathrm{Soc}(P) \longrightarrow 0 .$$

In particular, if $M = k$, then $X \cong P_k \oplus H(P_k)$ where $H(P_k) = \mathrm{Rad}(P_k)/\mathrm{Soc}(P_k)$ is the *heart of* the projective cover of the trivial $kG$-module $k$ (cf. [5]).

The main result regarding endotrivial $kG$-modules is [13, Theorem 2.6]. For completeness, the Dynkin diagrams mentioned in the theorem are the following (cf. [10, Sect. 2.30] for a full list):

$$A_n = \circ \underline{\qquad} \circ \underline{\qquad} \ldots \underline{\qquad} \circ \quad (n \text{ vertices})$$

$$\tilde{A}_{1,2} = \circ \overset{(2,2)}{\underline{\qquad}} \circ$$

$$\tilde{B}_n = \circ \overset{(1,2)}{\underline{\qquad}} \circ \underline{\qquad} \circ \cdots \circ \underline{\qquad} \circ \overset{(2,1)}{\underline{\qquad}} \circ \quad (n+1 \text{ vertices})$$

$$A_\infty = \circ \underline{\qquad} \circ \underline{\qquad} \circ \ldots$$

$$A_\infty^\infty = \cdots \circ \underline{\qquad} \circ \underline{\qquad} \circ \ldots$$

$$D_\infty = \circ$$

**Theorem 2.11.** *Let $V$ be an indecomposable endotrivial $kG$-module and suppose that $\Gamma$ is the component of the stable Auslander–Reiten quiver containing $V$. Let $S \in \mathrm{Syl}_p(G)$.*

1. *If $S$ is cyclic, then $\Gamma$ is a Dynkin diagram of type $A_n$ and the endotrivial modules are exactly the modules that lie in the two end orbits.*
2. *If $S \cong C_2 \times C_2$ and $N_G(S) = C_G(S)$, then $\Gamma = \tilde{A}_{1,2}$ and all the modules in $\Gamma$ are endotrivial.*

3. If $S \cong C_2 \times C_2$, if $N_G(S) \neq C_G(S)$, and if $k$ does not contain a primitive cube root of 1, then $\Gamma = \tilde{B}_3$ and the endotrivial modules are exactly the modules that lie in the two end orbits.
4. If $S$ is dihedral and the cases 2 and 3 do not hold, then $\Gamma = A_\infty^\infty$ and all the modules in $\Gamma$ are endotrivial.
5. If $S$ is semidihedral, then $\Gamma = D_\infty$ and the endotrivial modules are exactly the modules that lie in the two end orbits.
6. In all the other cases, $\Gamma = A_\infty$ and the endotrivial modules form the unique end orbit.

## 2.5 Simple Endotrivial Modules

An $RG$-module $M$ is *simple*, or *irreducible*, if $M \neq \{0\}$ and the only submodules of $M$ are $\{0\}$ and $M$ itself. In [88, 89, 105, 110, 111], the focus has been on the study of simple endotrivial modules, likely inspired by [38, Conjecture 3.6].

Robinson proves in [110] that the study of simple endotrivial $kG$-modules that are not monomial may be reduced to the case when $G$ is quasi-simple. Recall that a $kG$-module $M$ is *monomial* if its matrix representation is a monomial matrix, or equivalently, if there exists a basis $\{m_1, \ldots, m_t\}$ of $M$ such that $gm_i = a_{ji}(g)m_j$ for all $1 \leq i, j \leq t$ and for some $a_{ji}(g) \in k$ depending on $g$ and $i$ (cf. [50, Sects. 43 and 52]). This work builds on joint work of Navarro and Robinson, who prove the following in [105, Theorem].

**Theorem 2.12.** *Suppose that $G$ is $p$-soluble of $p$-rank at least 2 and let $V$ be a simple endotrivial $kG$-module. Then $\dim(V) = 1$.*

Their proof uses the fact that the outer automorphism group of a finite simple group has a Sylow $p$-subgroup of order at most $p$, a fact which relies on the classification of finite simple groups. They proceed by contradiction on a minimal counter-example, minimal with respect to the dimension of $V$ and then minimal with respect to the order of $G$. In particular, such a module $V$ must be faithful as a $kS$-module, for $S \in \mathrm{Syl}_p(G)$, implying that $O_p(G) = 1$. Recall that an $RG$-module $V$ is *faithful* if its *kernel* $\ker(V) = \{g \in G \mid gv = v , \, \forall v \in V\}$ is trivial. Hence [105, Corollary] shows that $V$ must be *primitive*, that is, $V$ is not induced from a proper subgroup of $G$. Navarro and Robinson then apply what they call "Clifford-theoretic reductions" to obtain a contradiction, starting with a minimal normal subgroup $K$ of $G$ containing $Z(G)$, and working in $G/Z(G)$. Since $O_p(G) = 1$, the quotient $K/Z(G)$ is a $p'$-group, and the minimality of $K$ says that $K/Z(G)$ is a direct product of simple groups. They then observe that $V\downarrow_K^G$ remains irreducible and that, either $K$ is nilpotent, or it is a central product of the centre $Z(G)$ of $G$ with a unique conjugacy class of components of $G$. A *component* of $G$ is a quasi-simple subnormal subgroup of $G$, where a subgroup $H$ of $G$ is *subnormal* if there exists a chain $H = H_0 \trianglelefteq H_1 \trianglelefteq \cdots \trianglelefteq H_m = G$ of subgroups of $G$ such that $H_j \trianglelefteq H_{j+1}$ for all $0 \leq j < m$, written $H \triangleleft \triangleleft G$.

By minimality, it follows that $G$ must be a semi-direct product $K \rtimes Q$ where $Q$ is an elementary abelian $p$-subgroup of $G$ of rank 2. They then deduce that there exists a $x \in Q$ such that $C_K(x)$ is perfect and such that the nontrivial simple $kC_K(x)$-modules have dimension greater than 1. The required contradiction follows by Green's correspondence and an application of *Brauer's third main theorem* ([11, Vol I, Theorem 6.4.5]):

**Theorem 2.13** (Brauer's third main theorem). *Let $H$ be a subgroup of $G$ containing $DC_G(D)$ for some $p$-subgroup $D$ of $G$ and $b$ a block of $kH$ with defect group $D$. Then the Brauer correspondent $b^G$ of $b$ is the principal block of $kG$ if and only if $b$ is the principal block of $kH$.*

Following-up on the case when $G$ is $p$-soluble, [110, 111] state several reductions of the study of simple endotrivial modules. When $G$ has $p$-rank at least 2 and there exists a simple endotrivial $kG$-module, Robinson observes that if there exists a subnormal subgroup $N$ of $G$ of order divisible by $p$, then $V\downarrow_N^G$ is simple (cf. [110, Lemma 2]). From which he deduces that if $N$ is a component of $G$ and $V$ is faithful with $\dim(V) > 1$, then $N$ must be the only component of $G$. The main theorem of the paper shows the following.

**Theorem 2.14.** *Let $G$ be a finite group of $p$-rank at least 2 and let $V$ be a faithful simple endotrivial $kG$-module. Then either the product of the components is a quasi-simple subgroup of $G$ acting irreducibly on $V$, or else $O_{p'}(G)$ is abelian self-centralising and $V = U\uparrow_H^G$ for some 1-dimensional $kH$-module $U$ and some strongly $p$-embedded subgroup $H$ of $G$.*

Elaborating further on the theme, and considering simple endotrivial $\mathscr{O}G$-modules, [111] aims to characterise some properties of groups possessing such representations. The proofs are based on arguments involving a substantial theoretical knowledge in group theory, far beyond the scope of the present text, and therefore we only state the main results, referring the interested reader to the article and the specialised literature.

**Theorem 2.15.** *Let $G$ be a finite group of $p$-rank at least 2.*

1. *([111, Corollary 2.4]) If $V$ is a faithful simple endotrivial $\mathscr{O}G$-module of rank greater than 1 and if $N = O_p(G) > 1$, then $C_G(N) = N \times Z(G)$ with $N$ elementary abelian and $V$ of $\mathscr{O}$-rank equal to the $R$-rank of $N$ minus 1. Moreover, either $N \in \mathrm{Syl}_p(G)$, or $N \cong C_2 \times C_2$ and $G \cong Z(G) \times \mathfrak{S}_4$.*
2. *([111, Lemma 2.5]) If $V = W\uparrow_H^G$ for some $H < G$, then $W$ is simple and $H$ must be strongly $p$-embedded in $G$.*
3. *([111, Lemma 2.6]) If $V$ is primitive simple, then $O_{p'}(G) \le Z(G)$.*
4. *([111, Lemma 2.9]) If $V$ is a simple faithful endotrivial $\mathscr{O}G$-module, then $V\downarrow_{O^p(G)}^G$ is irreducible.*
5. *([111, Theorem 3.2]) If $V$ is a simple faithful endotrivial $\mathscr{O}G$-module and if $S \in \mathrm{Syl}_p(G)$ has no cyclic subgroup of index 2, then $G$ has a unique conjugacy class of involutions (i.e. elements of $G$ of order 2).*

In particular, Robinson proves that for $p = 2$, if $|G|$ is even and if $V$ is a simple faithful endotrivial $\mathcal{O}G$-module of rank greater than 1, then $O^2(G)$ has a unique conjugacy class of involutions. If $V$ is imprimitive, then $G$ must have some strongly 2-embedded subgroup and a unique conjugacy class of involutions, while if $V$ is primitive and $O_2(G) > 1$, then either $O_2(G) \in \mathrm{Syl}_2(G)$ is elementary abelian, or $O_2(G) \cong C_2 \times C_2$ and $G \cong Z(G) \times \mathfrak{S}_4$, otherwise $O_2(G) = 1$ and $O_{2'}(G) \leq Z(G)$.

The subsequent papers [88, 89] introduce a character-theoretic approach to the investigation of simple endotrivial modules, in particular the simple faithful trivial source endotrivial modules. We will describe their method and present some applications in Sect. 5.1. In addition, the authors study the properties of a block $B$ of $kG$ which would contain some indecomposable endotrivial $kG$-module in the case when $G$ has cyclic Sylow $p$-subgroups. Their results build on those of [13], described in Sect. 2.4 above, and those of [104], which we will review in Sect. 3.6. The key point is that if $S \in \mathrm{Syl}_p(G)$ is cyclic and if $Z$ denotes the subgroup of $S$ of order $p$, then $H = N_G(Z)$ is strongly $p$-embedded in $G$ and has a nontrivial normal $p$-subgroup. So any trivial source endotrivial $kG$-module $V$ is isomorphic to the $kG$-Green correspondent of a 1-dimensional $kH$-module. Hence, in [89, Lemma 3.1], they prove that the inertial index of $B$ is equal to the inertial index $|H : C_G(Z)|$ of the principal block. In other words, $B$ contains as many nonisomorphic simple $kG$-modules as the principal block. Thus $B$ contains $2e$ indecomposable endotrivial modules, which are exactly the modules forming the two boundary orbits of the stable Auslander–Reiten quiver of $B$: one orbit consisting of all $\Omega_G^{2j-1}(V)$ and the other of all $\Omega_G^{2j}(V)$, for $1 \leq j \leq e$. In particular, all the modules in one boundary have the same dimension as $k$-vector spaces.

They then show, in [89, Lemma 3.2], that $B$ contains an indecomposable endotrivial $kG$-module $V$ if and only if its Brauer correspondent $b$ contains a 1-dimensional $kH$-module, and $kG$ has precisely $\dfrac{|H|}{|H'S|e}$ such blocks, each of which contains at least one simple endotrivial module. Furthermore, $V$ must be isomorphic to the $kG$-Green correspondent of $\Omega_H^j(U)$ for some 1-dimensional $kH$-module $U$.

We refer to [1] for a concise reference for the Brauer tree. This is a visual representation of the projective covers of the simple modules in a block of $kG$ with cyclic defect group. Briefly, the *Brauer tree* is a finite tree, i.e. a simply laced undirected connected graph without loops or cycles, together with:

1. a *cyclic ordering* of the edges at each vertex, and
2. a *multiplicity* attached to one of the vertices, namely a positive integer. This vertex is then called the *exceptional vertex*.

The convention is to draw a Brauer tree $\mathcal{T}$ following an anti-clockwise numbering of the edges emanating from a given vertex (note that each edge has two labels since it joins two vertices). An edge is a *leaf* if it is the unique edge of a vertex at one of its ends.

The edges in $\mathcal{T}$ are in bijection with the isomorphism classes of simple modules in a given block of $kG$. Now, if $V$ is simple and $P$ is its projective cover, we know from [1, Theorem 22.1] that $\mathrm{Rad}(P)/V \cong U \oplus W$ is the direct sum of two uniserial modules $U$ and $W$, possibly zero. The Brauer tree helps us to read off the composition series of $U$ and $W$. For instance, if $V$ appears in a portion of $\mathcal{T}$ of the form

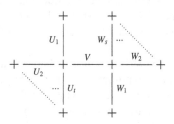

where no end vertex of $V$ is the exceptional one, then one of the uniserial direct summands of $P$, say $U$, has composition factors $U_1, U_2, \ldots, U_t$ in that order, from top to bottom, and similarly for $W$. If one of the ends of $V$ is the exceptional vertex with multiplicity, say $m$, then we "walk around" the exceptional vertex $m$ times listing the modules, so that $U$ would then have composition factors $U_1, \ldots, U_t, V, U_1 \ldots, U_t, V, U_1 \ldots, U_t$, where the sequence of $U_j$'s occurs $m$-times. Brauer trees for the blocks of finite group algebras with cyclic defect groups have long been studied and are generally well known. What [89] adds to the theory pertaining to Brauer trees is the location of the endotrivial modules in a Brauer tree of a block $B$ of $kG$ with full defect in the case when the group $G$ has cyclic Sylow $p$-subgroups. Their main result is that if $B$ contains an endotrivial module and $H \neq C_G(Z)$, then a simple $kG$-module $V$ in $B$ is endotrivial if and only if $V$ corresponds to a leaf in $\mathcal{T}$ which does not join the exceptional vertex ([89, Theorem 3.7]). Hence they apply their result to $\mathrm{SL}_2(q)$, for $q \equiv \pm 1 \pmod{p}$, and $V$ a simple $kG$-module. They show in [89, Proposition 3.8] that $V$ is endotrivial if and only if one of the following holds:

1. $p$ is odd and divides $q - 1$, and $V$ lies in a $p$-block with full defect (which is cyclic) and inertial index 2, or
2. $p$ and $q$ are odd and $p$ divides $q + 1$, and $V$ lies in the non-principal block with full defect (which is cyclic) and inertial index 2, or
3. $p = 3$ divides $q + 1$ and $|S| = 3$, and $V$ lies in the principal block (which is cyclic).

Moreover, if $p = 2$ and $q \equiv -1 \pmod 4$, and if $V$ lies in the principal block, then $V$ is endotrivial as a $k\,\mathrm{PSL}_2(q)$-module but not as a $k\,\mathrm{SL}_2(q)$-module.

Needless to say, the proofs require a thorough knowledge of the blocks and Brauer trees for $\mathrm{SL}_2(q)$.

# Chapter 3
# Classifying Endotrivial Modules

This chapter aims to present the results on the classification of endotrivial modules which have been obtained using traditional representation theory methods. By *classification*, we mean a description of the abelian group structure of $T(G)$, and if known, a presentation of $T(G)$ by generators and relations.

## 3.1 "Classifying" Endotrivial Modules

From Dade's early findings on endotrivial modules, it seems reasonable to believe that endotrivial modules can be "classified". That is, we can describe the indecomposable endotrivial $RG$-modules up to isomorphism for any finite group $G$ and $R = \mathcal{O}$ or $R = k$ in any suitable $p$-modular system $(F, \mathcal{O}, k)$. In [53], Dade classifies the endotrivial $RG$-modules for $G$ an abelian $p$-group.

The first step towards the classification of endotrivial modules is Dade's theorem [53, Theorem 10.1]:

**Theorem 3.1.** *Suppose that $G$ is a finite abelian p-group. Then $T(G) = \langle \Omega_G \rangle$ is cyclic. More precisely, if $|G| \leq 2$, then $T(G)$ is trivial, if $G$ is cyclic of order at least 3, then $T(G)$ has order 2, and $T_R(G)$ is infinite cyclic otherwise.*

Building on these results and on Theorem 2.8, Puig marks a key step in the study of endo-permutation and endotrivial modules with the result on the finite generation of the Dade group [107, Corollary 2.4].

**Theorem 2.3.** *Let $G$ be a finite p-group and let $R = \mathcal{O}$ or $R = k$. Then the group $D(G)$ of capped endo-permutation $RG$-modules is finitely generated. In particular, the group $T(G)$ of endotrivial $RG$-modules is finitely generated.*

This result is in fact a corollary of the main result in [107], which asserts that the kernel of the restriction map $T(G) \to \prod_E T(E)$ is finite, where $E$ runs through

© The Author(s), under exclusive license to Springer Nature Switzerland AG 2019
N. Mazza, *Endotrivial Modules*, SpringerBriefs in Mathematics,
https://doi.org/10.1007/978-3-030-18156-7_3

the nontrivial elementary abelian subgroups of $G$. The above statement then follows from an argument using the deflation maps $D(G) \to D(N_G(Q)/Q)$, for a nontrivial subgroup $Q$ of $G$, and by induction on the order of $G$, given that Dade proved that the statement holds in the case of an abelian $p$-group $G$.

As a consequence, $T(G)$ can be written as the direct sum of its torsion subgroup $TT(G)$ and a finitely generated torsionfree direct sum complement $TF(G)$, so that

$$T(G) = TT(G) \oplus TF(G).$$

Note that the torsionfree complement $TF(G)$ is not unique, but its rank, as a torsion-free abelian group, is unique. By decomposing $T(G)$ into two parts, we can split the problem of finding $T(G)$ into two as well. That is, on the one side, we can investigate $TT(G)$, i.e. look for all the stable isomorphism classes of "torsion" endotrivial modules, and on the other side, we can try and find a set of generators for a suitable complement $TF(G)$; or at least, find the torsionfree rank of $T(G)$, by which we mean the rank of the free $\mathbb{Z}$-module $T(G)/TT(G)$.

Let us now raise the question of rationality for all of $T(G)$: *Given a p-modular system $(F, \mathcal{O}, k)$, what, if any, are the differences between the groups of $\mathcal{O}G$- and $kG$-endotrivial modules? Does any endotrivial $kG$-module lift to an endotrivial $\mathcal{O}G$-module?* The answer is given by Lassueur, Malle and Schulte in [89, Theorem 1.3] for arbitrary finite groups, elaborating on Alperin's argument in [4, Theorem], in the case of endotrivial modules for finite $p$-groups. Recall, from Sect. 1.3, that we assume throughout that our $\mathcal{O}G$-modules are $\mathcal{O}$-free (in the literature, such modules are often called $\mathcal{O}G$-lattices).

We say that a $kG$-module $M$ *lifts* to an $\mathcal{O}G$-module if there exists an $\mathcal{O}G$-module $L$ such that $M \cong L/\pi L$ as $kG$-modules, where $\pi$ is the unique maximal ideal of $\mathcal{O}$ (cf. Definition 1.1). We refer the reader to [85, Sect. I.14] for more details on existence and uniqueness of lifts.

**Theorem 3.2.** *Let $(F, \mathcal{O}, k)$ be a p-modular system and $G$ a finite group. An endotrivial $kG$-module lifts to an endotrivial $\mathcal{O}G$-module.*

*Proof.* (*Outline of proof*). Let us identify an endotrivial module $M$ with its matrix representation $\rho : G \to \mathrm{GL}_n(k)$, where $n = \dim_k(M)$, and replace $G$ with its image $\rho(G) \leq \mathrm{GL}_n(k)$. Then, working within the general linear group, we can show that $G$ has a normal subgroup $G_0$ of index prime to $p$ whose image lies in $\mathrm{SL}_n(k)$. As a consequence, it suffices to prove the theorem assuming that the image of $G$ as a subgroup of $\mathrm{GL}_n(k)$ lies in $\mathrm{SL}_n(k)$. Now, if $\rho(G) \leq \mathrm{SL}_n(k)$, then we consider the *congruence subgroups* $\mathrm{SL}_n(\mathcal{O}, m) = \{I_n + \pi^m A \mid A \in \mathrm{Mat}_n(\mathcal{O})\}$, where $\mathrm{Mat}_n(R)$ denotes the ring of $n \times n$ matrices with coefficients in $R$. The congruence subgroups are normal in $\mathrm{SL}_n(\mathcal{O})$, and they form a central series of $\mathrm{SL}_n(\mathcal{O}, 1)$. Moreover, the successive quotient groups $\mathrm{SL}_n(\mathcal{O}, m)/\mathrm{SL}_n(\mathcal{O}, m + 1)$ are all isomorphic to the $\mathrm{SL}_n(k)$-module $\mathfrak{sl}_n(k)$ of trace zero matrices, with respect to the action by conjugation of $\mathrm{SL}_n(k)$. By assumption $\mathrm{End}_k M \cong k \oplus (\mathrm{proj})$, where (proj) is the kernel of the trace map. Therefore $\mathfrak{sl}_n(k)$ is a projective $kG$-module, hence injective too. We now construct the pullback

$$
\begin{array}{ccc}
X_2 & \cdots\cdots\cdots\cdots\cdots\cdots\rightarrow & G \\
\vdots & & \downarrow{\scriptstyle \rho} \\
\end{array}
$$

$$
0 \longrightarrow \mathrm{SL}_n(\mathcal{O}, 1)/\mathrm{SL}_n(\mathcal{O}, 2) \longrightarrow \mathrm{SL}_n(\mathcal{O})/\mathrm{SL}_n(\mathcal{O}, 2) \xrightarrow{\bmod \pi} \mathrm{SL}_n(k) \longrightarrow 0,
$$

where the bottom sequence is exact, and where $\mathrm{SL}_n(\mathcal{O}, 1)/\mathrm{SL}_n(\mathcal{O}, 2) \cong \mathfrak{sl}_n(k)$ is injective. So, the bottom sequence splits, yielding a map $\rho_2 : G \to \mathrm{SL}_n(\mathcal{O})/\mathrm{SL}_n(\mathcal{O}, 2)$ with

$$
\ker(\rho_2) = \mathrm{SL}_n(\mathcal{O}, 1)/\mathrm{SL}_n(\mathcal{O}, 2) \cong \mathfrak{sl}_n(k).
$$

Inductively iterating the pullback construction for $m > 2$, we obtain group homomorphisms $\rho_m : G \to \mathrm{SL}_n(\mathcal{O})/\mathrm{SL}_n(\mathcal{O}, m)$ lifting $\rho_{m-1}$. As a result, we get a group homomorphism $\tilde{\rho} : G \to \mathrm{SL}_n(\mathcal{O})$ lifting $\rho$ by the universal property of projective limits. In other words, we obtain an $\mathcal{O}G$-module lifting $M$, corresponding to the representation $\tilde{\rho}$. Now if $L$ is a lift of $M$, we need to show that $L$ is endotrivial as an $\mathcal{O}G$-module. We know that the $\mathcal{O}$-rank of $L$ is equal to $\dim_k M$, and so congruent to $\pm 1 \pmod{p}$. It follows that $\mathrm{End}_{\mathcal{O}} L \cong \mathcal{O} \oplus V$, where $V$ must be projective because its reduction mod $\pi$ is projective as a $kG$-module. Therefore, $L$ is an endotrivial $\mathcal{O}G$-module lifting $M$. $\qquad\square$

Theorem 3.2 allows us to assume that $R = k$ from now on, which simplifies our task of classifying endotrivial modules over $kG$ only. Henceforth, unless otherwise specified, we write $T(G) = T_k(G)$.

## 3.2 Maximal Elementary Abelian Subgroups of a Finite Group and $TF(G)$

We start by discussing the torsionfree rank of $T(G)$. It turns out that the structure of the poset of elementary abelian $p$-subgroups of $G$ is crucial in finding the rank of $TF(G)$, as shown in the results in [3, Theorem 4], [43, Theorem 12.1] and [34, Proposition 2.3, Corollary 2.4].

**Theorem 3.3.** *Let $G$ be a finite group. The kernel of the restriction map*

$$
\mathrm{Res} : T(G) \longrightarrow \prod_{E \in \mathscr{E}_{\geq 2}(G)/G} T(E) \quad \text{is the torsion subgroup of } T(G),
$$

*where $\mathscr{E}_{\geq 2}(G)/G$ denotes the poset of $G$-conjugacy classes of elementary abelian $p$-subgroups of $G$ of rank at least 2. Hence $T(G)$ is a finitely generated abelian subgroup with torsion-free rank equal to $n$, where $n$ is the number of connected components of $\mathscr{E}_{\geq 2}(G)/G$.*

In particular, if $G$ is abelian, then $TF(G) = \mathbb{Z}$ if $G$ is noncyclic and is 0 if $G$ is cyclic (as we have seen in Theorem 3.1).

Let us describe the noncyclic elementary abelian $p$-subgroup structure of $G$, focussing on the poset (or category) $\mathscr{E}_{\geq 2}(G)$ of Theorem 3.3. This is the poset whose elements are the elementary abelian $p$-subgroups of $G$ of rank at least 2, with order relation given by the inclusion of subgroups. The group $G$ acts by conjugation on the elements of $\mathscr{E}_{\geq 2}(G)$, and we let $\mathscr{E}_{\geq 2}(G)/G$ be the poset of $G$-orbits, with the induced order relation

$$[E] \leq [F] \text{ in } \mathscr{E}_{\geq 2}(G)/G \iff E \leq_G F \text{ in } \mathscr{E}_{\geq 2}(G), \quad \text{for all } [E], [F] \in \mathscr{E}_{\geq 2}(G)/G.$$

We say that two elements $[E], [F] \in \mathscr{E}_{\geq 2}(G)/G$ are *connected* if there exist $p$-subgroups $E_1, \ldots, E_l$ in $\mathscr{E}_{\geq 2}(G)$ such that $E \leq_G E_1 \geq_G \cdots \geq_G E_l \leq_G F$. Thus the *connected components* of $\mathscr{E}_{\geq 2}(G)/G$ are the largest disjoint subsets of $\mathscr{E}_{\geq 2}(G)/G$ such that any two elements in one subset are connected and no two elements of distinct subsets are connected. A connected component reduced to a single conjugacy class of maximal elementary abelian $p$-subgroups (of rank 2) is called an *isolated vertex*.

*Example 3.1.*

1. If $G$ is a noncyclic abelian $p$-group, then $\mathscr{E}_{\geq 2}(G)/G$ is connected.
2. If $G$ is a finite group and $S \in \mathrm{Syl}_p(G)$ has a noncyclic centre $Z(S)$, then $\mathscr{E}_{\geq 2}(G)/G$ is connected.
3. If $p = 3$ and $G = C_3 \wr C_3$, then $G$ has rank 3 and $\mathscr{E}_{\geq 2}(G)/G$ can be represented by the graph

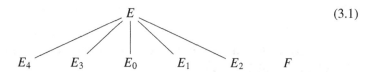

$$(3.1)$$

where $E$ is elementary abelian of rank 3, $E_0 = Z_2(S)$ is a normal elementary abelian subgroup of $G$ of rank 2 and $E_1, \ldots, E_4, F$ are elementary abelian of rank 2 and nonnormal in $G$. In this case, there are two connected components: an isolated vertex $\{[F]\}$ and a *big* component $\{[E], [E_j] \mid 0 \leq j \leq 4\}$, containing any subgroup of rank greater than 2 and a normal elementary abelian subgroup of rank 2.
4. If $G = \mathrm{SL}_3(p)$, then $S \cong p_+^{1+2}$ has rank 2 and $\mathscr{E}_{\geq 2}(S)/S \cong \mathscr{E}_{\geq 2}(S)$ consists of $p + 1$ isolated vertices if $p$ is odd, or 2 isolated vertices if $p = 2$. The poset $\mathscr{E}_{\geq 2}(G)/G$ consists then of 3, respectively 2 isolated vertices if $p$ is odd, respectively $p = 2$.

For a finite $p$-group $S$, the basic properties of $\mathscr{E}_{\geq 2}(S)/S$ are described in [67, Sect. 10] and in [43, Lemmas 2.1 and 2.2], from which we gather the following.

**Proposition 3.1.** *Let $G$ be a finite group and $S \in \mathrm{Syl}_p(G)$.*

1. *$\mathscr{E}_{\geq 2}(G)/G$ is nonempty unless $S$ is cyclic or quaternion (if $p = 2$). In particular, if $p$ is odd, then $\mathscr{E}_{\geq 2}(G)/G$ is nonempty if and only if $S$ has a normal elementary abelian subgroup of rank 2.*
2. *If $\mathscr{E}_{\geq 2}(G)/G$ has more than one connected component, then $Z(S)$ is cyclic.*
3. *If $S$ has rank 2, then $\mathscr{E}_{\geq 2}(G)/G$ is totally disconnected, that is, each connected component consists of a unique conjugacy class of elementary abelian p-subgroups of $G$ of rank 2.*
4. *If $S$ has rank at least 2, then $\mathscr{E}_{\geq 2}(G)/G$ contains a unique maximal connected component $\mathscr{B}$, containing all the conjugacy classes of elementary abelian p-subgroups of rank at least 3 and all the other components are isolated vertices. Moreover, $\mathscr{B}$ contains a normal elementary abelian p-subgroup of rank 2, unless $p = 2$ and $S$ is dihedral of order at least 16.*
5. *Suppose that $\mathscr{E}_{\geq 2}(G)/G$ is disconnected and let $F$ be a maximal elementary abelian subgroup of $S$ of rank 2. Then*

   - *$F = \langle u \rangle \times Z$, where $u$ is a noncentral element of $S$ of order $p$, and $Z$ the unique central subgroup of $S$ of order $p$.*
   - *$C_S(F) = C_S(u) = \langle u \rangle \times Q$, where $Q$ is cyclic, or possibly quaternion if $p = 2$.*
   - *$N_S(F) = C_S(F) * E_0$ is a central product over $Z$, where $E_0$ is an elementary abelian subgroup of $S$ of rank 2 and $F * E_0 \cong p_+^{1+2}$. Moreover, $E_0$ is normal in $S$, unless $p = 2$ and $S$ is dihedral of order at least 16.*

In particular, if $p$ is odd, if $S$ has rank at least 3 and if $\mathscr{E}_{\geq 2}(G)/G$ is disconnected, then the centraliser $C_S(F)$ in $S$ of a maximal elementary abelian subgroup $F$ of $S$ of rank 2 (necessarily nonnormal in $S$) is a noncyclic abelian metacyclic p-group, and $F$ is the unique noncyclic elementary abelian subgroup of $C_S(F)$.

*Remark 3.1.* The number of connected components of $\mathscr{E}_{\geq 2}(G)/G$ can also be defined as the number of $G$-conjugacy classes of maximal elementary abelian p-subgroups of rank 2 if $G$ has $p$-rank 2, or this number plus one if the $p$-rank of $G$ is greater than 2.

Suppose that $\mathscr{E}_{\geq 2}(G)/G$ has at least 2 components. Let $F$ be a nonnormal maximal elementary abelian subgroup of $S \in \mathrm{Syl}_p(G)$ of rank 2. It follows from Héthelyi's work in [78, 79] that the subgroup $C_S(F)$ is *soft* in $S$ whenever $C_S(F)$ is abelian.

**Definition 3.1.** Let $S$ be a finite p-group. A subgroup $A$ of $S$ is *soft* if $A = C_S(A)$ has index $p$ in its normaliser. We call $A$ *deep soft* if $|S : A| > p$, i.e. if $A$ is not normal in $S$.

We refer the reader to Héthelyi's articles and references therein for the numerous properties of p-groups with soft subgroups, and return instead to some of their consequences when applied in the special case of the presence of soft metacyclic p-subgroups for $p$ odd in [14, 64, 102]. For completeness, we include the related results for $p = 2$ in [28, 96].

**Theorem 3.4.** *Suppose that $\mathscr{E}_{\geq 2}(G)/G$ has at least 2 components. The following hold.*

1. *The $p$-rank of $G$ is at most $p$ if $p$ is odd, and at most 4 if $p = 2$.*
2. *$\mathscr{E}_{\geq 2}(G)/G$ has at most $p + 1$ connected components if $p$ is odd, and at most 5 if $p = 2$. Furthermore, both bounds are attained.*
3. *If $p$ is odd and $G$ has rank at least 3, then $G$ contains a unique normal elementary abelian subgroup $E$ of rank 2. In particular, $E$ is characteristic in $G$.*

Note that if $G$ has a normal elementary abelian subgroup $E$ of rank greater than $p$, then $\mathscr{E}_{\geq 2}(G)/G$ must be connected. Indeed, if $x \in G$ has order $p$, then conjugation by $x$ induces an automorphism of the $\mathbb{F}_p$-vector space $E$ (written additively) which has dimension greater than $p$. Since $x$ has order $p$, it acts by a matrix (in a given basis of $E$) with Jordan blocks of size at most $p$. Therefore, there must be at least two distinct Jordan blocks, and so two linearly independent eigenvectors. Observe that an "eigenvector" is an element in the centraliser of $x$, which says in this case that $C_E(x)$ has rank at least 2. So if $x \in E$, then $C_G(x) \geq C_E(x) = E \in \mathscr{E}_{\geq 2}$ has rank greater than $p$, and if $x \notin E$, then $\langle x \rangle \times C_E(x) \leq C_G(x)$ is elementary abelian of rank at least 3. We conclude that if $G$ has a normal elementary abelian subgroup of rank at least $p + 1$, then there cannot be maximal elementary abelian subgroups of rank 2, and consequently $\mathscr{E}_{\geq 2}(G)/G$ is connected. The tedious part of the proof is to show that this remains true whenever $G$ has rank greater than $p$ for $p$ odd, or 4 for $p = 2$.

Here is a direct consequence of Theorem 3.4 for endotrivial modules.

**Theorem 3.5.** *Suppose that $G$ is a finite group of $p$-rank $m_p$ at least 2.*

1. *$TF(G)$ is cyclic whenever $m_p > p$ if $p$ is odd or $m_p > 4$ if $p = 2$.*
2. *$TF(G)$ has rank at most $p + 1$ if $p$ is odd, and at most 5 if $p = 2$.*

Note that all these bounds are optimal.

- $C_p \wr C_p$ has rank $p$ and $T(C_p \wr C_p) = TF(C_p \wr C_p) = \mathbb{Z}^2$, for any prime $p$.
- $p_+^{1+2}$ has rank 2 and $TF(p_+^{1+2}) \cong \mathbb{Z}^{p+1}$ if $p$ is odd, and $TF(D_8) \cong \mathbb{Z}^2$.
- $Q_8 * D_8$ has rank 4 and $T(Q_8 * D_8) = \mathbb{Z}^5$.

## 3.3  Endotrivial Modules for Finite $p$-Groups

In this section we classify endotrivial modules for finite $p$-groups, following the steps of Carlson and Thévenaz in their three collaborations [42–44]. See also [29] for an account.

We begin with their initial conjecture about the *detection* of endotrivial modules. By detection, we mean a family of subgroups $\mathscr{H}$ of $G$ such that the restriction map $\mathrm{Res}_{\mathscr{H}}: T(G) \to \prod_{h \in \mathscr{H}} T(H)$ is injective. (In other words, if an indecomposable endotrivial module $M$ restricts to $k \oplus (\mathrm{proj})$ for each $H \in \mathscr{H}$, then $M = k$.)

**[42, Conjecture 2.6].** *Let $G$ be a noncyclic finite $p$-group. Then the restriction map* Res : $T(G) \to \prod_E T(E)$ *is injective, where $E$ runs over all subgroups of $G$ which are either elementary abelian of rank 2 or, in addition if $p = 2$, isomorphic to $Q_8$.*

Let $G$ be a finite noncyclic $p$-group. By [42, Theorem 2.7], the restriction map Res : $T(G) \to \prod_H T(H)$ is injective, where $H$ runs over all subgroups of $G$ that are:

- if $p$ is odd: either elementary abelian of rank 2, or extraspecial of exponent $p$;
- if $p = 2$: either elementary abelian of rank 2, almost extraspecial, or extraspecial except $D_8$ (cf. Sect. 1.2).

So [42, Conjecture 2.6] says that the (almost) extraspecial $p$-groups, except $Q_8$, are not needed for the detection of endotrivial modules. Note that $Q_8$ is indeed necessary, because the only elementary abelian 2-subgroup of a quaternion group is the centre $Z$, of order 2, which is not in the above list. (Also, $M{\downarrow}_Z^{Q_{2^n}} \cong k \oplus (\text{proj})$ for any endotrivial $kQ_{2^n}$-module $M$.) Therefore, for a quaternion group $Q_{2^n}$, we need to look at the restriction of modules to the smallest quaternion groups that are contained in $Q_{2^n}$.

The conjecture turns out to be true.

**[44, Theorem 1.4].** *If $G$ is extraspecial of exponent $p$ if $p$ is odd, or, if $p = 2$, (almost) extraspecial but not $D_8$, then the restriction map* Res : $T(G) \to \prod_H T(H)$ *is injective, where $H$ runs through all the maximal subgroups of $G$.*

The proof is tedious and requires a considerable background in group cohomology and support varieties for modules. The strategy consists in bounding the dimension of a critical module as a $k$-vector space, and show that the trivial module is the only critical module. A *critical module* is an indecomposable endotrivial $kG$-module $M$ whose restriction to all proper subgroups of $G$ is isomorphic to $k \oplus (\text{proj})$. The existence of a bound on $\dim_k(M)$ has long been known, but not its value. Carlson and Thévenaz estimate the bound on the dimension of a critical module for an almost extraspecial $p$-group of the type specified in [42, Theorem 2.7]. Here is a brief breakdown of the steps of the proof.

- If $G$ is an extraspecial $p$-group, then $\Phi(G) = Z \leq Z(G)$ and $\bar{G} := G/Z$ is elementary abelian of even rank, say $2r$. Suppose (by contradiction) that there exists a nontrivial critical module $M$. So $M \not\cong k \oplus (\text{proj})$ as $kG$-modules. Put

$$Z = \langle z \rangle \quad \text{and} \quad \bar{M} = M/M', \quad \text{where} \quad M' = \{x \in M \mid (z-1)^{p-1}x = 0\}.$$

  Then $\bar{M} \cong M'$ as $kG$-modules, and the quotient module $\bar{M}$ is a periodic $k\bar{G}$-module (note that $(z-1)M \subseteq M'$). Moreover, $\dim_k(M) = p\dim_k(\bar{M}) + 1$.
- Suppose that $\bar{M}$ splits as the direct sum $\bar{M} = \bar{M}_1 \oplus \bar{M}_2$ for $k\bar{G}$-modules $\bar{M}_1, \bar{M}_2$ such that the varieties of $\bar{M}_1$ and $\bar{M}_2$ intersect at $\{0\}$, then there exist critical $kG$-modules $N_1$ and $N_2$ such that $\bar{N}_i \cong \bar{M}_i$ for $i = 1, 2$, and $N_1 \otimes N_2 \cong M \oplus (\text{proj})$.
- Consequently, given critical modules $M_1, \ldots, M_n$ such that we can pick $n$ distinct lines $\ell_1, \ldots, \ell_n$ with $\ell_j \in V_{\bar{G}}(\bar{M}_j)$ for each $j$, we can construct a nontrivial critical module $M$ such that $V_{\bar{G}}(\bar{M}) = \cup_j \ell_j$ with $\dim_k(M) \geq \frac{n|G|}{2} + 1$ if $p = 2$ and $\dim_k(M) \geq n|G| + 1$ if $p$ is odd.

- Let $O_G$ be the symplectic group associated to the almost extraspecial $p$-group $G$. That is, if $p = 2$, then we have a quadratic form associated to $\bar{G}$, and if $p$ is odd, there is a symplectic form associated to $G$ (defined using commutators). It turns out that $O_G$ coincides with the group of outer automorphisms of $G$ fixing $Z(G)$. Moreover, there exists a critical $kG$-module $M$ such that

$$
\dim(M) > \begin{cases} |G| + \dfrac{|O_G|}{|C|} & \text{if is odd and} \\ \dfrac{|G|}{2} + \dfrac{|O_G|}{|C|} & \text{if } p = 2, \text{ where} \end{cases}
$$

  $C$ is a cyclic $p'$-subgroup of $O_G$ of maximal order. This provides a lower bound for the dimension of a critical module.

- Using cohomological arguments, we can show that for some positive integer $n$, we have for any integer $m$:

$$
\dim(\Omega_G^m(M)) + \dim(\Omega_G^{m-n+1}(M)) \le \sum_{1 \le j \le n} \dim(\Omega_G^{m-j+1}(k){\uparrow}_{H_j}^G),
$$

  where $H_1, \ldots, H_t$ are maximal subgroups of $G$ satisfying a certain vanishing product property in cohomology, with $n = t$ if $p = 2$, and $n = 2t$ if $p$ is odd. This provides an upper bound for the dimension of a critical module.

- Combining upper and lower bounds for the dimension of a putative critical module $M$, we carry out a case by case inspection of each isomorphism type of almost extraspecial group listed above: of exponent $p$ if $p$ is odd, or (almost) extraspecial except $D_8$ if $p = 2$. The upshot is that for each such $p$-group, assuming the existence of a nontrivial critical module leads to a contradiction. For instance, if $G = D_8 * D_8$, then we calculate $128 < \dim(M) \le 62$, which is absurd.

The conclusion drawn from the above argument is that critical modules do not exist and as a consequence [42, Conjecture 2.6] holds. Here is perhaps an easier way to summarise the result (cf. [43, Corollary 12.6]).

**Theorem 3.6.** *Let $G$ be a finite $p$-group. Then $TT(G) = 0$, unless $G$ is cyclic of order at least 3, quaternion or semi-dihedral.*

Once the torsion subgroup $TT(G)$ is determined for all finite $p$-groups $G$, it remains to find generators for a suitable $TF(G)$, in order to complete the classification of endotrivial modules for finite $p$-groups. We know that we can always choose such a set of generators containing $\Omega_G$, and that $TF(G) = \langle \Omega_G \rangle \cong \mathbb{Z}$ whenever $\mathscr{E}_{\ge 2}(G)/G$ is connected.

So, let $G$ be a nonabelian finite $p$-group such that $\mathscr{E}_{\ge 2}(G)/G$ has at least two connected components, and if $p = 2$, suppose that $G$ is not dihedral, a case that we postpone to Sect. 3.8. Assume the same notation as in Sect. 3.2, and in particular, Proposition 3.1. Let $E_0, \ldots, E_n$ be representatives of the $G$-orbits in $\mathscr{E}_{\ge 2}(G)/G$ with $E_0$ a normal elementary abelian subgroup of $G$ of rank 2, and $E_i = Z \times \langle u_i \rangle$ for a noncentral subgroup $\langle u_i \rangle$ of $G$ of order $p$, for $1 \leq i \leq p$. Then $C_G(\langle u_i \rangle)$ is of the form $\langle u_i \rangle \times Q_i$, where $Q_i$ is cyclic, or possibly quaternion if $p = 2$.

**Definition 3.2.** For each $1 \leq i \leq n$, let $N_i$ be the cap of

$$
\begin{cases}
\left(\Omega_G^{-1}(\Omega_{G/\langle u_i \rangle}(k))\right)^{\otimes 2} & \text{if } C_G(\langle u_i \rangle)/\langle u_i \rangle \text{ is cyclic of order } \geq 3, \\
\Omega_G^{-1}(\Omega_{G/\langle u_i \rangle}(k)) & \text{if } p = 2 \text{ and } |C_G(\langle u_i \rangle)/\langle u_i \rangle| = 2, \\
\left(\Omega_G^{-1}(\Omega_{G/\langle u_i \rangle}(k))\right)^{\otimes 4} & \text{if } p = 2 \text{ and } C_G(\langle u_i \rangle)/\langle u_i \rangle \text{ is quaternion.}
\end{cases}
$$

In other words, in the Dade group $D(G)$ of $G$, we have $[N_i] = \alpha(\Omega_{G/\langle u_i \rangle} - \Omega_G)$ for $\alpha = 1, 2$ or $4$, depending on $C_G(\langle u_i \rangle)/\langle u_i \rangle$.

(Here, we use that $T(G) = \bigcap_{1 < H < G} \ker\left(\mathrm{Defres}_{N_G(H)/xH}^G : D(G) \longrightarrow D(N_G(H)/H)\right)$.)

The key property of these modules is stated in [44, Theorem 3.1].

**Proposition 3.2.** *The modules $N_i$ are endotrivial, for $1 \leq i \leq n$. Furthermore*

$$
\mathrm{Res}_{E_j}^G N_i \cong
\begin{cases}
k \oplus (\mathrm{proj}) & \text{if } i \neq j, \\
\Omega_{E_j}^{-2p}(k) \oplus (\mathrm{proj}) & \text{if } i = j \text{ and } C_G(\langle u_i \rangle)/\langle u_i \rangle \text{ is cyclic of order } \geq 3, \\
\Omega_{E_j}^{-2}(k) \oplus (\mathrm{proj}) & \text{if } i = j \text{ and } |C_G(\langle u_i \rangle)/\langle u_i \rangle| = 2, \\
\Omega_{E_j}^{-8}(k) \oplus (\mathrm{proj}) & \text{if } i = j \text{ and } C_G(\langle u_i \rangle)/\langle u_i \rangle \text{ is quaternion.}
\end{cases}
$$

If we assume that $\mathcal{E}_{\geq 2}(G)/G$ has at least two connected components, then $TT(G) = \{0\}$ by Theorem 3.6, and so $T(G) = TF(G) \cong \mathbb{Z}^n$. Now, given that the restrictions of the modules $N_i$ to non-conjugate elementary abelian subgroups of $G$ are $\mathbb{Z}$-linearly independent and since the restriction map $\mathrm{Res} : T(G) \to \prod_j T(E_j)$ is injective, we obtain the main result on $TF(G)$ ([44, Theorem 7.1]).

**Theorem 3.7.** *Let $G$ be a nonabelian finite $p$-group, and if $p = 2$, suppose that $G$ is not dihedral, semi-dihedral or generalised quaternion. Then $TT(G) = 0$, and*

$$
TF(G) = \langle \Omega_G, N_1, \ldots, N_n \rangle \cong
\begin{cases}
\mathbb{Z}^n & \text{if } G \text{ has rank at most 2}; \\
\mathbb{Z}^{n+1} & \text{if } G \text{ has rank at least 3.}
\end{cases}
$$

*These generators are subject to one relation, of the form*

$$
\alpha \Omega_G + \sum_{1 \leq j \leq n} [N_j] \quad \text{for } \alpha \text{ as in Definition 3.2.}
$$

The 2-groups of maximal nilpotency class are handled separately in [42], on a case by case basis, by finding explicit endotrivial modules, and showing that the stable isomorphism classes generate all of the group of endotrivial modules. The main result is the following.

**Theorem 3.8.** *Let $G$ be a nonabelian finite 2-group.*

1. *([42, Theorem 6.3]) If k contains a primitive cube root of 1 and G is quaternion of order 8, then $T(G) = \langle \Omega_G, [\Omega_G(M)] \rangle \cong \mathbb{Z}/4 \oplus \mathbb{Z}/2$, where $\Omega_G(M)$ is a selfdual endotrivial module of dimension 5. If k does not contain a primitive cube root of 1, then $T(G) = \langle \Omega_G \rangle \cong \mathbb{Z}/4$.*
2. *([42, Theorem 6.5]) If G is quaternion of order at least 16, then $T(G) = \langle \Omega_G, [\Omega_G(M)] \rangle \cong \mathbb{Z}/4 \oplus \mathbb{Z}/2$, where M is an indecomposable endotrivial module with $\dim(M) = \frac{|G|}{2} - 1$, and $\mathrm{Res}_C^G(M) = \Omega_C(k)$ for the unique maximal cyclic subgroup C of G.*
3. *([42, Theorem 5.4]) If G is dihedral, then $T(G) = \langle \Omega_G, \Omega_{G/C} \rangle \cong \mathbb{Z}^2$, where C is a noncentral subgroup of G of order 2.*
4. *([42, Theorem 7.1]) If G is semi-dihedral, then $T(G) = \langle \Omega_G, [\Omega_G(\Omega_{G/C})] \rangle \cong \mathbb{Z} \oplus \mathbb{Z}/2$, where C is a noncentral subgroup of G of order 2.*

The two most important facts about $T(G)$ for a finite p-group G are that:

- $TT(G) = 0$ except if G is cyclic of order at least 3, semi-dihedral or quaternion.
- All the endotrivial modules are defined over the prime field $\mathbb{F}_p$, since they are stably isomorphic to modules built from relative syzygies, except if G is quaternion of order 8 and k contains a primitive cube root of 1. If $G = Q_8$, then there exists a selfdual endotrivial module which is not defined over $\mathbb{F}_2$, but is defined over any field extension of $\mathbb{F}_4$ (cf. [42, Theorem 6.3]).

## 3.4   Endotrivial Modules in the Normal Case

Dade defines endotrivial modules only for finite p-groups, being motivated by the study of endo-permutation modules, but these representations are also important in the representation theory of any finite group. Indeed, prior to Dade, Alperin, who was interested in the stable module category of a finite group, introduced the notion of *"invertible modules"* for any finite group algebra (cf. [2]), and this is what we now call an *endotrivial module*. So, following the completion of the classification for finite p-groups, the objective is now to find a classification of endotrivial modules for arbitrary finite groups. At present, this is still an open problem, and the methods of investigation have greatly diversified, as we shall see later.

In this section, we use the results presented in Sects. 2.1 and 2.2 to classify the endotrivial modules in the case when G is a finite group with a normal Sylow p-subgroup.

We start with a few generalities, proved in [34, Sects. 2, 3], which rely on traditional representation theory facts, such as the theory of vertices and sources, Green's correspondence, Mackey's formula and Clifford theory (cf. Sect. 1.1).

**Proposition 3.3.** *Let G be a finite group and $S \in \mathrm{Syl}_p(G)$. Let $N = N_G(S)$.*

1. *$T(G)$ is finitely generated. Hence $T(G) = TT(G) \oplus TF(G)$, where $TT(G)$ has finite order and $TF(G)$ is a finitely generated torsionfree subgroup of $T(G)$.*

2. *The restriction map* $\mathrm{Res}_N^G : T(G) \to T(N)$ *is injective.*
3. *If $H$ is a strongly $p$-embedded subgroup of $G$, then $\mathrm{Res}_H^G : T(G) \to T(H)$ is an isomorphism, with inverse $\mathrm{Ind}_H^G$. In particular, if $S$ is a trivial intersection subgroup of $G$, then $\mathrm{Res}_N^G$ is an isomorphism.*
4. *If $M$ is an endotrivial $kG$-module, then $M$ is a direct summand of $\mathrm{Ind}_S^G(V)$ for some endotrivial $kS$-module $V$.*
5. *Suppose that $M$ is a $kG$-module such that $\mathrm{Res}_S^G M$ is endotrivial. Then $M$ is endotrivial.*
6. *Suppose that $G = N$. Then an indecomposable $kG$-module $M$ is endotrivial if and only if $\mathrm{Res}_S^G M$ is an indecomposable endotrivial $kS$-module. In particular, if $G = N$ and if $S$ is not cyclic, quaternion or semi-dihedral, then $TT(G) \cong \mathrm{Hom}(G, k^\times)$ is isomorphic to the group of $1$-dimensional $kG$-modules.*
7. *The rank of $TF(G)$ as a free $\mathbb{Z}$-module is equal to the number of connected components of $\mathcal{E}_{\geq 2}(G)/G$.*

Suppose that $G$ has a normal Sylow $p$-subgroup $S$. As a corollary of Proposition 3.3, the restriction to $S$ of a minimal projective resolution of $k$ as a $kG$-module is a minimal projective resolution of $k$ as a $kS$-module [101, Corollary 2.7].

Given a $kG$-module, then its restriction to $S$ is a $kS$-module which is $G$-*stable*, that is, $M \cong {}^gM$ for all $g \in G$, where ${}^gM$ is the conjugate $kS$-module, as defined in Sect. 1.1. Conversely however, given a $G$-stable $kS$-module $M$, there is no $kG$-module $\tilde{M}$ which *extends* $M$, that is, such that $\tilde{M}\downarrow_S^G \cong M$. If $M$ is a $G$-stable endotrivial $kS$-module, then there is a key result due to E. Dade, in [54, Theorem 7.1]; a result that has never been published in this form (as far as we are aware).

**Theorem 3.9.** *Let $G$ be a finite group having a normal Sylow $p$-subgroup $S$, let $k$ be an algebraically closed field of characteristic $p$, and let $M$ be an endo-permutation $kS$-module. Then $M$ extends to a $kG$-module if and only if $M$ is $G$-stable.*

As an immediate consequence, we deduce the following.

**Corollary 3.1.** *The map $\mathrm{Res}_S^G : T(G) \to T(S)$ induces an isomorphism of abelian groups $TF(G) \cong TF(S)^{G/S}$, where $TF(S)^{G/S}$ is the subgroup of $TF(S)$ generated by the classes of the $G$-stable endotrivial $kS$-modules.*

Using the presentation of $T(S)$ by generators and relations given in Theorems 3.7 and 3.8, it comes down to a routine computation to obtain a presentation of $T(G)$ when $S \trianglelefteq G$. The following is [101, Lemma 3.5, Proposition 3.7].

**Proposition 3.4.** *Suppose $G$ has a normal Sylow $p$-subgroup $S$. Let $N_1, \ldots, N_n$ be the $kS$-modules as in Definition 3.2 and $E_1 \ldots, E_n$ the corresponding elementary abelian $p$-subgroups of $G$. That is, $E_i = Z \times \langle u_i \rangle$ and $N_i$ is constructed using $\Omega_{S/\langle u_i \rangle}$.*

1. $N_i \cong {}^gN_i$ *if and only if $g \in SN_G(E_i)$.*
2. *Each $N_i$ extends to an indecomposable endotrivial $k[SN_G(E_i)]$-module $\tilde{N}_i$.*
3. *The tensor induced module $M_i = \mathrm{Ten}_{SN_G(E_i)}^G \tilde{N}_i$ is endotrivial.*

The proof of the last assertion uses that $SN_G(\langle u_i \rangle)$ is the stabiliser of the $S$-conjugacy class of the group $\langle u_i \rangle$ (of order $p$). So by Dade's result, we can extend $\Omega_{S/\langle u_i \rangle}$ to $SN_G(\langle u_i \rangle)$, obtaining an indecomposable endotrivial $k[SN_G(\langle u_i \rangle)]$-module $M_i$ for each $i$. Then, a routine computation using Mackey's formula for tensor induction (cf. Sect. 1.3.4) shows that $\mathrm{Res}_{E_j}^{SN_G(\langle u_i \rangle)} M_i$ is endotrivial for all $1 \leq i, j \leq n$.

Combining Propositions 3.3 and 3.4, together with [39, 104], we obtain.

**Theorem 3.10.** *Let $G$ be a finite group having a normal Sylow $p$-subgroup $S$. Consider the same notation as above, and write the group $T(G)$ of endotrivial $kG$-modules as a direct sum $T(G) = TT(G) \oplus TF(G)$, where $TT(G)$ is the torsion subgroup and $TF(G)$ is torsionfree.*

*1. There is an exact sequence*

$$0 \longrightarrow \mathrm{Hom}(G, k^\times) \longrightarrow T(G) \longrightarrow T(S)^{G/S} \longrightarrow 0 .$$

*2. The set*

$$\left\{ \Omega_G, [\mathrm{Ten}_{PN_G(E_i)}^G \widetilde{N}_i] \mid 1 \leq i \leq n \right\}$$

*is a basis for $TF(G)$.*
*3. If $TT(S) = 0$, then $TT(G) \cong \mathrm{Hom}(G, k^\times)$ is isomorphic to the group of 1-dimensional $kG$-modules.*
*4. If $S$ is cyclic of order at least 3, then $T(S)^{G/S} = T(S) = \langle \Omega_G \rangle \cong \mathbb{Z}/2$.*
*5. If $S$ is quaternion or semi-dihedral, then $T(S)^{G/S} = T(S)$.*

There are (at least) two other constructions for a generating set of $TF(G)$, both using cohomological methods. We refer the reader to [101] for the full details on these alternative constructions. The outline is as follows. Let $G$ be a finite group with a normal Sylow $p$-subgroup $S$. Without loss, we assume that $\mathscr{E}_{\geq 2}(G)/G$ has at least two components, represented by $E_1, \ldots, E_n$, and say that $E_1$ is normal in $G$. Write each $E_j$ as $Z \times \langle u_j \rangle$, where $Z$ is the unique central subgroup of $S$ of order $p$. Thus $Z \triangleleft G$ and we put $\bar{\ } : G \to \bar{G} = G/Z$. Hence if $M$ is a $kG$-module, we let $M' = (z - 1)^{p-1} M$ and $\bar{M} = M/M'$, which we regard as a $k\bar{G}$-module.

The first method is coined the *deconstruction method* by Carlson and Thévenaz in [43]. We will come back to it in Sect. 3.10 below. To present this method, we start with an essential cohomological concept (see [26, Sect. 10]).

**Definition 3.3.** Let $G$ be a finite group and $k$ a field of positive characteristic $p$ with $p$ dividing $|G|$. The cohomology ring $\mathrm{H}^*(G, k)$ is a finitely generated, graded commutative $k$-algebra and has a maximal ideal spectrum $V_G(k)$ which is a homogeneous affine variety. If $M$ is a finitely generated $kG$-module, then $\mathrm{Ext}_{kG}^*(M, M)$ is a finitely generated module over $\mathrm{H}^*(G, k)$, and the *support variety* of $M$ is the set $V_G(M) = V_G(J(M)) \subseteq V_G(k)$, where $J(M)$ is the annihilator of $M$ in $\mathrm{H}^*(G, k)$ and $V_G(M)$ is the set of all maximal ideals that contain $J(M)$. So $V_G(M)$ is a closed homogeneous subvariety of $V_G(k)$. For $E \leq G$, let $\mathrm{res}_{G,E}^* : V_E(k) \longrightarrow V_G(k)$ denote the restriction map induced on the varieties by the inclusion of $E$ in $G$.

Here is an omnibus theorem about support varieties for modules which gives the properties of support varieties that we need (cf. [11, Vol II, Sects. 5.7 and 5.9]).

**Theorem 3.11.** *Suppose that $L$, $M$ and $N$ are $kG$-modules.*

1. *The module $M$ is projective if and only if $V_G(M) = \{0\}$.*
2. *$V_G(M^*) = V_G(\Omega_G^n(M)) = V_G(M)$ for any $n \in \mathbb{Z}$.*
3. *$V_G(M \oplus N) = V_G(M) \cup V_G(N)$, and $V_G(M \otimes N) = V_G(M) \cap V_G(N)$.*
4. *If the sequence $0 \to L \to M \to N \to 0$ is exact, then $V_G(M) \subseteq V_G(L) \cup V_G(N)$.*
5. *(Quillen's theorem) $V_G(M) = \cup_E res^*_{G,E}(V_E(M))$, where $E$ runs through all the elementary abelian $p$-subgroups of $G$.*
6. *If $V_G(M) = V_1 \sqcup V_2$, where $V_1$ and $V_2$ are nonempty closed sets. Then $M \cong M_1 \oplus M_2$, where $V_G(M_1) = V_1$ and $V_G(M_2) = V_2$.*
7. *Let $\hat{\zeta} : \Omega_G^n(k) \to k$ be a cocycle representing an element $\zeta \in \mathrm{H}^n(G, k)$, and let $L_\zeta = \ker(\hat{\zeta})$. Then $V_G(L_\zeta) = V_G(\zeta)$.*

Hence, let us now come back to the study of $T(G)$ with $S \trianglelefteq G$. Consider the same $\alpha_j$ as in Definition 3.2, i.e. $\alpha_j = 1$, 2 or 4, depending on whether $C_S(\langle u_j \rangle)/\langle u_j \rangle$ has order 2, is cyclic of order at least 3, or is quaternion. In [27], Carlson proves that the support variety of each $\overline{\Omega_G^{a_i}(k)}$ decomposes as the union of two nonempty closed subvarieties $V_{\bar{G}}(\overline{\Omega_G^{a_i}(k)}) = V_j \cup V_j'$ subject to $V_j \cap V_j' = \{0\}$, $V_j = res^*_{\bar{G}, \bar{E}_j}(V_{\bar{E}_j}(k))$ and $res^*_{\bar{G}, \bar{E}_i}(V_{\bar{E}_i}(k)) \subseteq V_j'$, for all $i \neq j$. Here, the restriction is the restriction map from Quillen's theorem (cf. [26, Theorem 10.1]). So if $\mathscr{I}$ is an ideal in $\mathrm{H}^*(E_j, k)$ belonging to $V(L)$ for some $kE_j$-module $L$, then its restriction is the maximal ideal $\{\zeta \in \mathrm{H}^*(G, k) \mid \mathrm{Res}^G_{E_j}(\zeta) \in \mathscr{I}\}$ of $\mathrm{H}^*(G, k)$. Thus, we obtain a map $res^*_{G, E_j} : \mathrm{Spec}(\mathrm{H}^*(E_j, k)) \longrightarrow \mathrm{Spec}(\mathrm{H}^*(G, k))$, where $\mathrm{Spec}(A)$ denotes the maximal ideal spectrum of the ring $A$. Then, [27, Theorem 4.2] shows that, for each $1 \leq j \leq n$, there exists an indecomposable endotrivial module $N_j$ such that $N_j \downarrow_{E_j}^G \cong \Omega_{E_j}^{\alpha_j}(k) \oplus (\mathrm{proj})$ and $N_j \downarrow_{E_i}^G \cong k \oplus (\mathrm{proj})$ for all $i \neq j$.

The argument is much more abstract than the previous one, but it has the advantage of "producing" indecomposable endotrivial $kG$-modules, though not in an explicit way.

The second cohomological method is called the *cohomological push-out method*, and like the previous "decomposition" method, the arguments invoked are highly technical, and the result not as explicit as in our first method.

Loosely, instead of dis-assembling modules, we build new endotrivial modules by "assembly". The outcome consists of "torsionfree" endotrivial $kG$-modules $M_i$ such that

$$M_i \downarrow_{E_i}^G = \Omega_{E_i}^{m_i}(k) \oplus (\mathrm{proj}) \quad \text{and} \quad M_i \downarrow_{E_j}^G \cong k \oplus (\mathrm{proj}) \quad \text{for all } 1 \leq i \neq j \leq n,$$

whenever $\mathrm{H}^{m_i}(G, k)$ contains an element $\zeta$ such that $\mathrm{Res}^G_Z(\zeta) \in \mathrm{H}^{m_i}(Z, k)$ is not nilpotent (i.e. nonzero, and if $p$ is odd, then $m_i$ is even). In particular, if we know that $m_i$ is the minimum positive degree such that the $n$-tuple

$$(0, \ldots, 0, m_i \Omega_{E_i}, 0 \ldots, 0) \in \prod_j T(E_j), \quad \text{where } \Omega_{E_i} \text{ is in the } i\text{th position,}$$

can possibly belong to the image of the restriction map

$$\mathrm{Res}_{\mathscr{E}_{\geq 2}(G)/G} : T(G) \to \prod_j T(E_j) \quad \text{for each } 1 \leq i \leq n,$$

then we have found a generating set $\{\Omega_G, [M_1], \ldots, [M_n]\}$ of $TF(G)$, subject to exactly one relation.

To construct the modules $M_i$, we use Carlson's $L_\zeta$ modules, defined for a non-nilpotent cohomology class $\zeta \in \mathrm{H}^m(G, k)$, for some positive integer $m$. "Explicitly", let $\tilde{\zeta} \in \mathrm{Hom}_{kG}(\Omega_G^m(k), k)$ be a cocycle representing $\zeta$ and $L_\zeta = \ker(\tilde{\zeta})$ (cf. [11, Sect. 5.9, Vol I]). Then, we know that the variety $V_G(\zeta) = V_G(L_\zeta)$ is disconnected. Namely, it is the disjoint union of all the $\mathrm{res}^*_{G, E_i}(V_{E_i}(k))$. As a consequence, $L_\zeta$ is a direct sum, say $L_\zeta \cong L_1 \oplus \cdots \oplus L_n$, of indecomposable $kG$-modules, one for each connected component of $\mathscr{E}_{\geq 2}(G)/G$. The method elaborated by Carlson in [27] yields the existence of an endotrivial module $N_i$ for each $1 \leq i \leq n$ as the push-out of the diagram

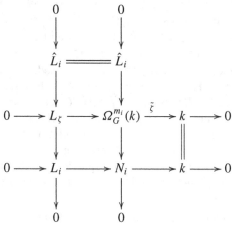

where $\hat{L}_i$ is the direct sum of all the summands of $L_\zeta$ except $L_i$. To see that $N_i$ is endotrivial, we calculate its restriction to an arbitrary noncyclic elementary abelian $p$-subgroup $E$ of $G$. By the properties of the summands of $L_\zeta$, the restriction of the diagram to $E$ are isomorphic to either $k \oplus (\mathrm{proj})$, or $\Omega_E^{m_i}(k)$, depending on whether we pick $E$ in the same component as the subgroup $E_i$ corresponding to $L_i$. Thus, the bottom row splits if we restrict the diagram to $E_j$ for $j \neq i$, because then $L_i{\downarrow}_{E_j}^G$ is projective, and it follows that $N_i{\downarrow}_{E_j}^G \cong k \oplus (\mathrm{proj})$. Instead, the middle column splits if we restrict to $E_i$, because then $\hat{L}_i$ is projective, and it follows that $N_i{\downarrow}_{E_i}^G \cong \Omega_{E_i}^{m_i}(k) \oplus (\mathrm{proj})$.

## 3.5  General Strategy

In this section, we lay the theoretical setting and describe the basic steps that we follow in order to determine $T(G)$ for an arbitrary finite group $G$.

Throughout, let $S$ be a Sylow $p$-subgroup of $G$ and write $N = N_G(S)$. In view of Sect. 3.4, we know the structure of $T(N)$, and so we may assume that $N$ is a proper subgroup of $G$. The traditional method used to compare the representation theory of $G$ and $N$ is the Green correspondence. In our case, Proposition 3.3 says that the restriction map $\mathrm{Res}_N^G : T(G) \to T(N)$ is injective, and that $\ker\left(\mathrm{Res}_S^N : T(N) \to T(S)\right) \cong \mathrm{Hom}(N, k^\times)$ is the group of stable isomorphism classes of the 1-dimensional $kN$-modules. Now, a 1-dimensional $kN$-module is a group homomorphism $\rho : N \to \mathrm{GL}_1(k) \cong k^\times$ which must have any commutator in its kernel and $S \leq \ker(\rho)$ too. So, by the isomorphism theorem, $\rho$ factors through $N/N'S$, and we see that the distinct group homomorphisms $N/N'S \to k^\times$ give nonequivalent representations of $N$ of degree 1. Therefore $\mathrm{Hom}(N, k^\times) \cong N/N'S$. Let us record an elementary fact.

**Lemma 3.1.** *With the above notation, $|G/G'S| \leq |N/N'S|$.*

*Proof.* The proof is a direct application of Frattini's argument (Sect. 1.2). Since $G'S \trianglelefteq G$, we have $G = (G'S)N$. It follows that $G/G'S \cong N/(N \cap G'S)$. Since $N'S \leq N \cap G'S$, the assertion holds.                                                                 $\square$

Now, as in Proposition 3.3, write $T(G) = TT(G) \oplus TF(G)$, where $TT(G)$ is the torsion subgroup of $T(G)$, which is finite, and $TF(G)$ is a finitely generated torsionfree subgroup of $T(G)$ of $\mathbb{Z}$-rank equal to the number of connected components of $\mathscr{E}_{\geq 2}(G)/G$. The main unknown in this "equation" is the following subgroup of $TT(G)$.

**Definition 3.4.** Let $G$ be a finite group having a nontrivial Sylow $p$-subgroup $S$. Define
$$K(G) = \ker\left(\mathrm{Res}_S^G : T(G) \to T(S)\right).$$

We call $K(G)$ the group of *trivial Sylow restriction $kG$-modules*. Note that $K(G)$ is a finite abelian $p'$-group containing the group of 1-dimensional $kG$-modules.

In other words, $[M] \in K(G)$ if and only if $M{\downarrow}_S^G \cong k \oplus (\mathrm{proj})$. Equivalently, $K(G)$ is the subgroup of $T(G)$ generated by the stable isomorphism classes of the indecomposable trivial source endotrivial $kG$-modules. In particular, any 1-dimensional $kG$-module is indecomposable trivial source and so $G/G'S \hookrightarrow K(G)$, via the identification of $G/G'S$ with the group of stable isomorphism classes of 1-dimensional $kG$-modules.

The restriction to $K(G)$ of the restriction map $\mathrm{Res}_N^G : T(G) \to T(N)$ induces an injective group homomorphism $\mathrm{Res}_N^G : K(G) \to K(N) = N/N'S$. So the upshot is that $G/G'S \hookrightarrow K(G) \hookrightarrow K(N)$, and between the two extremes, anything can happen. For instance, if $S \cap {}^gS \neq 1$ for all $g \in G$, then $G/G'S \cong K(G)$ (cf. Lemma 3.2

below), whilst if $S$ is a trivial intersection subgroup of $G$, then the restriction map $T(G) \to T(N)$ is an isomorphism and $K(G) \cong K(N)$, which can be much bigger than $G/G'S$.

Since little more can be said on $T(G)$ in general, the strategy consists in investigating special cases, such as $S$ cyclic, dihedral, quaternion or semidihedral, or specific classes of groups, which we do in Sects. 3.6 to 3.11.

**Lemma 3.2.** *Let $G$ be a finite group with a Sylow $p$-subgroup $S$ of order at least 3.*

1. *Suppose that $S \cap {}^g S \neq 1$ for all $g \in G$. Then, $K(G) \cong G/G'S$.*
2. *Suppose that $H$ is a normal subgroup of $G$ of index not divisible by $p$, and that $M$ is an indecomposable endotrivial $kG$-module. Then $M\downarrow_H^G$ is an indecomposable endotrivial $kH$-module. In particular, if $K(H) \cong H/H'S$, then $K(G) \cong G/G'S$.*
3. *Suppose that $S$ is cyclic or generalised quaternion, so that $Z(S)$ has a unique subgroup $Z$ of order $p$. Then $H = N_G(Z)$ has a nontrivial normal $p$-subgroup and $H$ is strongly $p$-embedded in $G$. Therefore,*

$$K(G) \cong K(H) \cong H/H'S, \quad and \quad \mathrm{Res}_H^G : T(G) \to T(H) \text{ is an isomorphism.}$$

*Proof.* 1. Let $[M] \in \ker(\mathrm{Res}_S^G)$. That is, $M\downarrow_S^G = k \oplus (\mathrm{proj})$ and $M \mid k\uparrow_S^G$. Using Mackey's formula, we obtain $M\downarrow_S^G \mid (k\uparrow_S^G)\downarrow_S^G \cong k^{|N_G(S):S|} \oplus \left( \bigoplus_x k\uparrow_{^xS \cap S}^S \right)$, where $x$ runs through a set of representatives of the double cosets $[S\backslash G/S]$ that are not in $N_G(S)$. Now $k\uparrow_{^xS \cap S}^S \cong k[S/({}^xS \cap S)]$ is an indecomposable permutation $kS$-module which is not projective because ${}^xS \cap S \neq 1$ for all $x \in G$. Therefore, since $M$ is endotrivial, we must have $M\downarrow_S^G \cong k$, i.e. $M$ has dimension 1.

2. Suppose that $M\downarrow_H^G \cong V \oplus (\mathrm{proj})$ for some indecomposable endotrivial $kH$-module $V$. Since $p$ does not divide $|G : H|$, the theory of vertices and sources (Sect. 1.4.3) applies and says that $M$ is a direct summand of $V\uparrow_H^G$. Therefore, $M\downarrow_H^G$ is a direct summand of

$$V\uparrow_H^G\downarrow_H^G \cong \bigoplus_{x \in [G/H]} {}^xV \quad (\text{cf. Sect. 1.3.1})$$

is a sum of conjugate modules because $H \trianglelefteq G$. But then the only way that $M\downarrow_H^G \cong V \oplus (\mathrm{proj})$ can be isomorphic to a direct summand of $\bigoplus_{x \in [G/H]} {}^xV$ is that $M\downarrow_H^G \cong V$ is indecomposable because $V$ is not projective. In particular, if $K(H) \cong H/H'S$ and $[M] \in K(G)$, then $M$ must have dimension 1.

3. We need to show that $p$ does not divide the order of $H \cap {}^xH$ for all $x \in G - H$. Suppose $H \cap {}^xH$ contains a nontrivial $p$-subgroup $Q$. By assumption on $S$, any $p$-subgroup of $H$ must contain $Z$, and $Z$ is the unique subgroup of $O_p(H)$ of order $p$. So if $1 < Q \leq H \cap {}^xH$, then $Z \leq Q \cap Q^x$ and $Z = Z^x$. But then $H = {}^xH$

because $x \in H$. Therefore $H$ is strongly $p$-embedded.

The second part of the claim is part of Proposition 3.3 and we now give a direct proof. Suppose that $M$ is an indecomposable endotrivial $kH$-module. Then $M\uparrow_H^G = V \oplus X$ and $V\downarrow_H^G = M \oplus Y$, where $V$ is the $kG$-Green correspondent of $M$, and where $X$ and $Y$ are direct sums of indecomposable modules whose vertices are conjugate to $p$-subgroups of $H \cap {}^xH$ for some elements $x \notin H$. Since $H$ is strongly $p$-embedded in $G$, intersections of distinct conjugates of $H$ do not have nontrivial $p$-subgroups, which means that $V\downarrow_H^G \cong M \oplus (\text{proj})$ is endotrivial. $\qquad \square$

In [92], the authors investigate the inflation map between $G$ and a quotient $\bar{G} = G/K$ of $G$, for a subgroup $K \le O_{p'}(G)$. We know that the inflation homomorphism $\mathrm{Inf}_{\bar{G}}^G : T(\bar{G}) \to T(G)$ is injective, but not surjective in general. For instance, the central extensions $2A_6$ and $3A_6$ of the alternating group $\bar{G} = A_6$ on 6 letters give $T(2A_6) \cong \mathbb{Z} \oplus \mathbb{Z}/8$, whilst $T(A_6) \cong \mathbb{Z} \oplus \mathbb{Z}/4$ in characteristic 3; and $T(3A_6) \cong \mathbb{Z}^2 \oplus \mathbb{Z}/3$, whilst $T(A_6) \cong \mathbb{Z}^2$ in characteristic 2. Lassueur and Thévenaz's analysis focusses on the comparison between $T(G)$ and $T(\tilde{G})$, where $\tilde{G}$ is a $p'$-representation group of $G/O_{p'}(G)$ in the case when $G$ has $p$-rank at least 2 and no strongly $p$-embedded subgroup (cf. Lemma 3.2). That is, $\tilde{G}$ is a central extension

$$1 \longrightarrow Z \longrightarrow \tilde{G} \xrightarrow{\pi} G \longrightarrow 1$$ of $G$, with $\tilde{G}$ of minimal order and such that for each $k$-vector space $V$, any group homomorphism $\theta : G \to \mathrm{PGL}(V)$ can be "lifted" to a group homomorphism $\tilde{\theta} : \tilde{G} \to \mathrm{GL}(V)$ such that the diagram

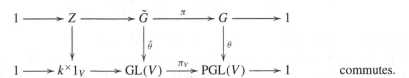

commutes.

Using the classification of finite simple groups, Lassueur and Thévenaz prove the following.

**Theorem 3.12.** ([92], Sects. 3 and 4) *Let $G$ be a finite group of $p$-rank at least 2 with no strongly $p$-embedded subgroup. Then there exists an injective group homomorphism $T(G)\big/(G/G'S) \to T(\tilde{G})\big/(\tilde{G}/\tilde{G}'S)$. In particular, if $K(\tilde{G}) \cong \tilde{G}/\tilde{G}'S$, then $K(G) \cong G/G'S$.*

*Moreover, there is always a group isomorphism $T(G) \cong T(G/[G, O_{p'}(G)])$.*

## 3.6 Cyclic Sylow $p$-Subgroups

In this section, suppose that $G$ has a cyclic Sylow $p$-subgroup $S$ and write $N = N_G(S)$. Let $Z$ be the unique subgroup of $S$ of order $p$ and let $H = N_G(Z)$. Theorem 3.10 deals with the case $N = G$. Since $T(S) = \langle \Omega_S \rangle \cong \mathbb{Z}/2$ if $|S| \geq 3$ and $0$ if $|S| \leq 2$, the restriction maps $T(G) \to T(S)$ and $T(H) \to T(S)$ are surjective. Lemma 3.2 says that the restriction $T(G) \to T(H)$ is an isomorphism because $H$ is strongly $p$-embedded in $G$, and therefore the sequence

$$0 \longrightarrow H/H'S \longrightarrow T(G) \longrightarrow T(S)^{H/S} \longrightarrow 0 \qquad (3.2)$$

is exact. To determine when the sequence splits, we need to look at $H/C_G(Z)$ as a subgroup of $\mathrm{Aut}(Z) \cong C_{p-1}$. Suppose that $H/C_G(Z)$ has order $e$, where $e$ must divide $p - 1$. We know that:

1. There exists a $\mu \in \mathrm{Hom}(H, \mathbb{F}_p^\times)$ of order $e$ and such that ${}^h u = u^{\mu(h)}$ for some $\mu(h) \in \mathbb{F}_p^\times$, for all $u \in Z$ and all $h \in H$. So $\mu \in \mathrm{Hom}(H, \mathbb{F}_p^\times) \cong H/H'S$ and $\ker(\mu) \geq C_G(Z)$. (Note that since $H/C_G(Z)$ is an abelian $p'$-group, we have $H'S \leq C_G(Z)$.)
2. If $|S| \geq 3$, then $\Omega_H(k)$ has order $2e$. That is, a projective resolution of $k$ as a $kH$-module is periodic of period $2e$.

Observe that $\mathrm{Res}_S^H(2\Omega_H) = 2\Omega_S = 0$, and so $2\Omega_H \in \ker(\mathrm{Res}_S^H)$. We also know that the composition factors of the projective cover $P_k$ of $k$ as a $kH$-module are, from top to bottom, $k, \mu, \mu^2 \ldots, \mu^{p-2}, k$. Moreover, $\Omega_H(k) = \mathrm{Rad}(P_k)$, whose projective cover $P_\mu$ is also uniserial, with composition factors from top to bottom $\mu, \mu^2 \ldots, k, \mu$. So, $2\Omega_H = [\Omega_H^2(k)]$ is the stable isomorphism class of the $kH$-module $\ker(P_\mu \to \mathrm{Rad}(P_k)) \cong \mu$. It follows that

$$T(G) \cong \langle H/H'S, \Omega_H \rangle \cong \big( H/H'S \oplus \Omega_H \big) / \langle [\mu] - 2\Omega_H \rangle,$$

and the sequence (3.2) splits whenever $\mu$ is a square in $\mathrm{Hom}(H, \mathbb{F}_p^\times)$.

Explicitly, the elements of $T(H)$ are of the form $[\chi \otimes \Omega_H^j(k)] = [\chi] + j\Omega_H$, for some $\chi \in \mathrm{Hom}(H, \mathbb{F}_p^\times)$ and $0 \leq j < e$. Since $H$ is strongly $p$-embedded in $G$, the elements of $T(G)$ are of the form $[\mathrm{Ind}_H^G(\chi \otimes \Omega_H^j(k))]$.

Note that if $p = 2$, then $H = C_G(Z)$ and $e = 1$. So the sequence (3.2) always splits, and more generally it splits whenever $e$ is odd.

## 3.7  Klein Four Sylow 2-Subgroups

Suppose that $G$ has a Klein four Sylow 2-subgroup $S = C_2 \times C_2$. Since $\mathrm{Aut}(S) \cong C_3$, we must have $N = N_G(S) = C_G(S)H$, with $|H| = 1$ or $3$, and $C_G(S) = S \times O(C_G(S))$, where $O(C_G(S)) = O_{2'}(C_G(S))$ is the largest normal subgroup of $C_G(S)$ of odd order. In [82], Koshitani and Lassueur investigate endotrivial modules and blocks with full defect of $kG$ in characteristic 2, assuming that $k$ is large enough for $G$. Their results give conditions for the restriction map $\mathrm{Res}_N^G : TT(G) \to TT(N)$ to be an isomorphism. As an unexpected outcome of their findings, it turns out that if a block with full defect contains a torsion endotrivial $kG$-module, then every $kG$-Green correspondent of a 1-dimensional $kN$-module that lies in that block is also endotrivial. The following is a compilation of Koshitani and Lassueur's results. Recall from [11, Vol I, Definition 6.2.3] that, given a $p$-subgroup $Q$ of $G$, the *Brauer homomorphism* is the ring homomorphism $\mathrm{Br}_Q^G : (kG)^Q \to kC_G(Q)$ defined by

$$\sum_{g\in G} \lambda_g g \mapsto \sum_{g\in C_G(Q)} \lambda_g g \ \ \text{for any} \ \ \sum_{g\in G} \lambda_g g \in (kG)^Q, \quad \text{where}$$

$(kG)^Q$ is the set of $Q$-fixed points of $kG$.

**Theorem 3.13.** [82] *Let $G$ be a finite group with a Klein four Sylow 2-subgroup $S$. Write $N = N_G(S)$. Let $V$ be a 1-dimensional $kN$-module belonging to the block $b = kNf$ of $kN$, and let $M$ be its $kG$-Green correspondent, belonging to the block $B = kGe$ of $kG$.*

1. *$M$ is endotrivial if and only the following two conditions hold:*

   a *$\mathrm{Br}_Q^G(e)$ is a block idempotent of $kC_G(Q)$ for any $p$-subgroup $Q$ of $S$; and*
   b *$\phi_u(1) = 1$ for each nontrivial element $u \in S$, where $\phi_u$ is the unique Brauer character ([11, Vol I, Sects. 5.3 and 6.2]) of a 2-block $b_u$ of $kC_G(u)$ with $b_u^G = B$.*

   *In particular, if $M$ is endotrivial, then all the $kG$-Green correspondents of the 1-dimensional $kN$-modules lying in $B$ are also endotrivial. Hence all the $kG$-Green correspondents of the 1-dimensional $kN$-modules lying in the principal block, or in any block containing a 1-dimensional $kG$-module, are endotrivial.*
2. *If $O(G)$ is abelian, then the restriction map $\mathrm{Res}_N^G : TT(G) \to TT(N)$ is an isomorphism.*

**Theorem 3.14.** [83] *Let $G$ be a finite group with a Klein four Sylow 2-subgroup $S$. Write $N = N_G(S)$ and $\bar{G} = G/O(G)$.*

1. *If $\bar{G} \cong S$ or $A_4$, then $K(G) \cong G/G'S$.*
2. *If $\bar{G} \cong A_5$, then $K(G) \cong K(H) \cong H/H'S$, where $H$ is a strongly 2-embedded subgroup of $G$ such that $H/O(H) \cong A_4$. Moreover, if $[M] \in K(G)$ with $M$ nontrivial indecomposable and belonging to the principal block of $kG$, then $\dim(M) = 5$.*
3. *If $\bar{G} \cong PSL_2(q) \rtimes C_n$ with $5 < q \equiv \pm 3$ (mod 8) and $n$ odd, then $K(G) \cong G/G'S \oplus \mathbb{Z}/3$. Moreover, if $[M] \in K(G)$ with $M$ nontrivial indecomposable and belonging to the principal block of $kG$, then $\dim(M) = \dfrac{q-1}{2}$ if $q \equiv 3$ (mod 8), or $\dim(V) = q$ if $q \equiv 5$ (mod 8).*

## 3.8   Dihedral Sylow 2-Subgroups

Suppose that $G$ has a dihedral Sylow 2-subgroup $S$ of order $2^n$ with $n \geq 3$. The main result on $T(G)$ shows that the restriction map $T(G) \to T(S)$ is split surjective. Let $N = N_G(S)$ and $Z = Z(S)$. Recall that $|Z| = 2$. It is well known that the automorphism group of a dihedral group is a 2-group ([65, Lemma 7.7.2]), and therefore the inclusion $N/C_G(S) \hookrightarrow \mathrm{Aut}(S)$ implies that $N = SC_G(S) = S \times K$, where $K = O(C_G(S)) = O(N) \cong C_G(S)/Z$ is the largest normal subgroup of $N$ (or equivalently of $C_G(S)$) of odd order. Preceeding by many years the classification of endotrivial modules for finite $p$-groups, Auslander and Carlson obtained a crucial result on endotrivial modules for finite groups with dihedral Sylow 2-subgroups.

**Proposition 3.5.** ([7] Lemma 5.4 and Proposition 5.5) *Let $G$ be a finite group. Let $P_k$ denote the projective cover of the trivial module, and let $E = \mathrm{Rad}(P_k)/\mathrm{Soc}(P_k)$ denote its heart.*

1. *$E$ is indecomposable unless $G$ has a dihedral Sylow 2-subgroup of order $2^n$ with $n \geq 3$, or $G$ has a Sylow 2-subgroup $Q = C_2 \times C_2$, and $k$ contains a primitive cube root of unity whenever $N_G(Q) \neq C_G(Q)$.*
2. *If $E$ is decomposable then it is the direct sum of two endotrivial modules which are the dual of each other.*

So, if $G = S$ is dihedral of order at least 8, then $T(G) = \langle \Omega_G, \Omega_{G/C} \rangle \cong \mathbb{Z}^2$ as stated in Theorem 3.8.

If $G > S$, Proposition 3.5 shows that

$$TF(G) = \langle \Omega_G, [M] \rangle \cong \mathbb{Z}^2,$$

where $\mathrm{Rad}(P_k)/\mathrm{Soc}(P_k) \cong M \oplus M^*$.

Then, restriction induces an injective group homomorphism between the torsion subgroups $TT(G) \to TT(N) \cong K/K'$, since $N = S \times K$. We then need to determine which $kG$-Green correspondents of the 1-dimensional $kN$-modules are endotrivial, in addition to the 1-dimensional $kG$-modules. If $K = 1$, or if $S$ is trivial intersection, or if $Z \leq Z(G)$, then the question is settled. More precisely, in [83, Theorem 1.2] Koshitani and Lassueur prove the following (using the classification of finite simple groups).

**Theorem 3.15.** *Let $G$ be a finite group with dihedral Sylow 2-subgroup $S$ of order at least 8, and write $\bar{G} = G/O(G)$. Suppose that $k$ is large enough. Then $K(G) \cong G/G'S$, unless possibly if $\bar{G} \cong A_6$, in which case it may be that every trivial Sylow restriction $kG$-module is a simple module of dimension 1 or 9, and, identifying $G/G'S$ with a subgroup of $K(G)$, the quotient group $K(G)/(G/G'S)$ is an elementary abelian 3-group.*

## 3.9 Semi-dihedral Sylow 2-Subgroups

Suppose that $G$ has a semi-dihedral Sylow 2-subgroup $S$ of order $2^n$ with $n \geq 4$. Let $N = N_G(S)$ and $Z = Z(S)$. Recall that $|Z| = 2$ and that $T(S) = \langle \Omega_S, [M] \rangle \cong \mathbb{Z} \oplus \mathbb{Z}/2$, where $M = \Omega_S(\Omega_{S/C}(k))$ for a noncentral subgroup $C$ of $S$ of order 2, by Theorem 3.8. By definition, $K(G)$ has odd order, and therefore [39, Proposition 6.1] yields

$$T(G) \cong K(G) \oplus \mathrm{im}\left( \mathrm{Res}_S^G : T(G) \to T(S) \right) \hookrightarrow N/N'S \oplus \mathbb{Z} \oplus \mathbb{Z}/2.$$

The question now is: *Is the restriction map surjective?* In other words, is there a $kG$-module $V$ such that $V{\downarrow}_S^G \cong \Omega_S(\Omega_{S/C}(k)) \oplus (\mathrm{proj})$. The answer is stated in [39, Corollary 6.3 and Theorem 6.4] and relies on work of Erdmann [57–59] and Webb [122] on the stable Auslander–Reiten quiver of $kG$ and almost split sequences (overlapping with Auslander–Carlson's result [7] mentioned above).

**Theorem 3.16.** *Let $G$ be a group with a semi-dihedral Sylow 2-subgroup $S$.*

1. *Let $P_k$ be the projective cover of $k$ and let $E = \mathrm{Rad}(P_k)/\mathrm{Soc}(P_k)$ be the heart of $P_k$. There is an almost split sequence of the form*

$$0 \longrightarrow V \longrightarrow E \oplus (\mathrm{proj}) \longrightarrow U \longrightarrow 0$$

   *where $V$ and $U$ are indecomposable modules in the component of the stable Auslander–Reiten quiver of $kG$ containing $\Omega_G(k)$ and $U$ is not isomorphic to $\Omega^{-1}(k)$. Moreover $V^* \cong U$.*
2. *The modules $V$ and $U$ are endotrivial.*
3. *$\Omega_G(U) \cong \Omega_G(U)^*$ and so $2[\Omega_G(U)] = 0$ in $T(G)$.*
4. *The restriction map $\mathrm{Res}_S^G : T(G) \longrightarrow T(S)$ is split surjective.*

At present, nothing more is known about the structure of $K(G)$ if $G$ has a semi-dihedral Sylow 2-subgroup $S$ in general, unless $S \cap {}^gS \neq 1$ for all $g \in G$, or unless $G$ has a strongly embedded 2-subgroup. There are a few specific instances for which we have determined $K(G)$, such as $G = \mathrm{GL}_n(q)$ with $q \equiv 3 \pmod 4$ (in which case we have $K(G) \cong G/G'S$, cf. [37, Theorem 10.1]). We will come back to these particular examples in Chap. 6.

## 3.10 Quaternion Sylow 2-Subgroups

Suppose that $G$ has a quaternion Sylow 2-subgroup $S$ of order $2^n$ with $n \geq 3$. Let $Z = Z(S)$, $N = N_G(S)$ and $H = C_G(Z)$. Then $H$ is strongly $p$-embedded in $G$, so that $\mathrm{Res}_H^G : T(G) \to T(H)$ is an isomorphism. Without loss, we can therefore suppose that $G = H$, i.e. $Z$ is central in $G$. In particular, $G$ has a nontrivial normal 2-subgroup and therefore $K(G) \cong G/G'S$ is isomorphic to the group of 1-dimensional

$kG$-modules. Since $T(S) = \langle \Omega_S, [U] \rangle \cong \mathbb{Z}/4 \oplus \mathbb{Z}/2$, for some selfdual endotrivial module $U$ of dimension $\dfrac{|S|}{2} + 1$ (assuming that $k$ contains a primitive cube root of 1 if $|S| = 8$), we have

$$T(G) \cong G/G'S \oplus \mathrm{im}\left(\mathrm{Res}_S^G : T(G) \to T(S)\right).$$

Note that $U$ is constructed differently for $|S| = 8$ and for $S$ of order at least 16. On the other hand, it is well known that a projective resolution of $k$ as a $kG$-module is periodic of period 4, independently of $|S|$, and so $\Omega_G$ has order 4 in $T(G)$ whenever $S$ is quaternion (cf. [39, Proposition 3.5]). The only open question is about the existence of an endotrivial $kG$-module $M$ such that $M{\downarrow}_S^G \cong U \oplus (\mathrm{proj})$.

In [26, Sect. 10], our argument uses cohomology and the fact that we assume $Z$ is normal in $G$. Let us give an outline of the proof, which is similar to the cohomological "deconstruction" method described in Sect. 3.4, and so we refer to Definition 3.3 and Theorem 3.11 (see also [26, Sect. 10]).

Write $\overline{H} = H/Z$ for any subgroup $H$ of $G$ containing $Z = Z(G)$ and $\overline{x} = xZ/Z \in \overline{G}$ for $x \in G$. Then $\overline{S}$ is dihedral, possibly Klein four.

Given a $kG$-module $M$, let $M_0 = \{v \in M \mid (z-1)v = 0\}$. Since $z^2 = 1$ and $k$ has characteristic 2, $M_0$ contains the submodule $(z-1)M$ of $M$. Moreover, $z$ acts trivially on $M_0$, and so $M_0$ is a $k\overline{G}$-module. Consider $W = \Omega_G^2(k)$. Since $W$ is endotrivial, $W{\downarrow}_Z^G \cong k \oplus (\mathrm{proj})$. So $(z-1)W = \mathrm{Soc}(\mathrm{proj})$ and $W_0 = k \oplus \mathrm{Soc}(\mathrm{proj})$, giving that $(z-1)W$ has codimension 1 in $W_0$. In other words, we have a filtration

$$\{0\} \subset (z-1)W \subset W_0 \subset W \quad \text{with} \quad W_0/(z-1)M = k \quad \text{and} \quad W/W_0 \cong (z-1)W.$$

If $G = S$, then $(z-1)W \cong W_1 \oplus W_2$ with both summands indecomposable of dimension $\frac{|S|}{4}$. Moreover, the variety of each $W_i$ regarded as a $k\overline{S}$-module is a distinct line (in different components of $W_{\overline{S}}(k)$ if $|S| \geq 16$).

From the case of a quaternion 2-group, we gather that for a finite group $G$ with quaternion Sylow 2-subgroup $S$ and $Z \lhd G$, if $M = \Omega_G^2(k)$, then (cf. [39, Proposition 3.4]):

$$(z-1)M \cong M_1 \oplus M_2 \quad \text{as} \quad k\overline{G}\text{-modules, with} \quad V_{\overline{G}}\big((z-1)M\big) = V_{\overline{G}}(M_1) \oplus V_{\overline{G}}(M_2),$$

and, for $i = 1, 2$, regarding $M_i$ as $kG$-modules, $M_i{\downarrow}_S^G \cong W_i \oplus (\mathrm{proj})$ for the modules $W_1, W_2$ above. Moreover, the modules $M_i$ are indecomposable of dimension congruent to $\frac{|S|}{4}$ (mod $\frac{|S|}{2}$). Because we assume that $Z$ is normal in $G$, hence central, the 1-dimensional quotient module $M_0/(z-1)M$ is trivial. Now, [39, Proposition 4.2] shows that, for $i = 1, 2$, there exist filtrations $\{0\} \subset X_i \subset Y_i \subset M$ of $M$, such that $(Y_i/X_i){\downarrow}_Z^G$ is endotrivial and $(z-1)(Y_i/X_i) \cong M_i$. It follows that the $kG$-module $Y_i/X_i$ is endotrivial and indecomposable for $i = 1, 2$. We then observe that $\dim(Y_i/X_i) = 1 + 2\dim(M_i) \equiv 1 + \frac{|S|}{2}$ (mod $|S|$), and so we obtain endotrivial modules $Y_i/X_i$ such that $(Y_1/X_1){\downarrow}_S^G \cong U \oplus (\mathrm{proj})$ (and $(Y_2/X_2){\downarrow}_S^G \cong \Omega_S^2(U)$, or vice versa).

## 3.11 Finite $p$-Soluble Groups

In this section, we describe $T(G)$ for $G$ a finite $p$-soluble group of $p$-rank at least 2, the cases of $p$-rank 1 being handled in Sects. 3.6 and 3.10. More precisely, our focus here is on finding $K(G)$, the group of trivial Sylow restriction $kG$-modules. In [38], we prove a partial result by induction on the $p$-soluble length of $G$ and state a conjecture which, if true, would give a complete description of $T(G)$ as an abelian group. The overarching result appears in [105], in which the outstanding conjecture of [38] is proved.

Suppose that $G$ is $p$-soluble (cf. Sect. 1.2). So either $O_p(G) > 1$, or $O_{p'}(G) > 1$. If $O_p(G) > 1$, then the group $K(G)$ is isomorphic to the group of 1-dimensional $kG$-modules. If $O_{p'}(G) > 1$, then there may be indecomposable trivial Sylow restriction $kG$-modules of dimension greater than 1.

Assume that the $p$-soluble length is 1, in the sense that either $G = S \rtimes K$, or $G = K \rtimes S$ with $S \in \mathrm{Syl}_p(G)$. In the first case, $S$ is normal in $G$ and we refer to Sect. 3.4. In the second case, $G$ is $p$-nilpotent.

There are several characterisations of finite $p$-nilpotent groups. The point is that such groups "resemble" finite $p$-groups, in the sense that if $G$ is $p$-nilpotent and $S \in \mathrm{Syl}_p(G)$, then $S$ controls the $p$-fusion in $G$, and each block of $kG$ is *nilpotent* as defined in [118, Sect. 49], which implies that each block contains a unique simple $kG$-module.

Suppose that $G = K \rtimes S$, with $K = O_{p'}(G)$ and $S \in \mathrm{Syl}_p(G)$, and let $V$ be a simple $kK$-module with inertia subgroup $H$ (cf. Sect. 1.3.4). Let $e$ be the central primitive idempotent of $kK$ corresponding to $V$ (cf. [118]). Then the following facts are well known [38, Lemma 3.1]:

1. $e$ is a block idempotent of $kH$ and $f = \sum_{g \in [G/H]} geg^{-1}$ is a block idempotent of $kG$. Moreover, induction induces a Morita equivalence between $\mathrm{mod}(kHe)$ and $\mathrm{mod}(kGf)$.
2. If $H < G$, then every module in $\mathrm{mod}(kGf)$ has dimension divisible by $p$.
3. If $H = G$, then $V$ can be equipped with the structure of a $kG$-module extending the given $kN$-module structure, and this extension is unique. Moreover, $V \downarrow_S^G$ is a capped endo-permutation $kS$-module, and $V$ is the unique simple module in the block $kGe$. For any $kGe$-module $Y$, the restriction $Y \downarrow_K^G$ is isomorphic to a direct sum of copies of the simple module $V \downarrow_K^G$. Finally, the module categories $\mathrm{mod}(kGe)$ and $\mathrm{mod}(kS)$ are Morita equivalent via the functor

$$\Phi : \mathrm{mod}(kS) \longrightarrow \mathrm{mod}(kGe), \quad X \mapsto V \otimes \mathrm{Inf}_S^G X,$$

where we identify $S$ with $G/K$ by a chosen isomorphism $S \xrightarrow{\cong} G/K$.

Suppose that $M$ is an indecomposable endotrivial $kG$-module, say $M$ belongs to the block $B$ of $kG$ (cf. Sect. 1.3.2).

Because we assume that $G$ is $p$-nilpotent, there exists a unique simple $kG$-module lying in $B$. Now, $\dim(M)$ is coprime to $p$, and so Item 2 above shows that the inertial group of a simple summand $V$ of the restriction $M{\downarrow}_K^G$ is equal to $G$. (Note that $kK$ is semisimple.) Therefore the $kK$-module $V$ has a unique extension to $G$, which must be the unique simple module in $B$ ([38, Corollary 3.2]).

**Theorem 3.17.** ([38] Theorem 3.3) *Let $G = K \rtimes S$ be a finite $p$-nilpotent group.*

1. *There is a split short exact sequence*

$$0 \longrightarrow K(G) \longrightarrow T(G) \xrightarrow{\text{Res}_S^G} T(S) \longrightarrow 0 ,$$

*where the splitting is obtained via $\text{Inf}_S^G$ and a chosen isomorphism $S \xrightarrow{\;\cong\;} G/K$. In other words,*

$$T(G) \cong K(G) \oplus T(S).$$

2. *Let $M$ be an indecomposable endotrivial $kG$-module. Then $[M] \in K(G)$ if and only if $M$ admits a decomposition $M \cong V \otimes U^*$, where $V$ is the unique simple $kG$-module in the block containing $M$ and $U$ is a $kS$-module which is a source for $V$, regarded as a $kG$-module via inflation $\text{Inf}_S^G$. Moreover, in that case, the simple module $V$ is endotrivial and its source $U$ is endotrivial.*

Theorem 3.17 raises the question of finding simple endotrivial $kG$-modules.

**Conjecture.** ([38] Conjecture 3.6) *Let $G = K \rtimes S$ be a finite $p$-nilpotent group with $S \in \text{Syl}_p(S)$ of $p$-rank at least 2.*

1. *If a simple $kG$-module $V$ is endotrivial, then $\dim(V) = 1$.*
2. *$K(G) \cong G/G'S \cong K/K'$.*

[38, Conjecture 3.6] was proved by Navarro and Robinson in [105, Theorem]. In fact their result is sharper as it also shows that there cannot be a simple "torsionfree" endotrivial module.

**Theorem** [105] *Let $G$ be a finite $p$-soluble group of $p$-rank at least 2. Let $V$ be a simple endotrivial $kG$-module. Then $V$ is 1-dimensional.*

In view of this result, [38, Theorem 6.2] becomes:

**Theorem 3.18.** *Let $G$ be a finite $p$-soluble group with $p$-rank at least 2. Then $K(G)$ is isomorphic to the group of 1-dimensional $kG$-modules. This is the case in particular if $G$ is soluble.*

Consequently, the only remaining open question about $T(G)$ for $G$ a finite $p$-soluble group of $p$-rank at least 2 is about generators for $TF(G)$ in the case when $TF(G)$ is not cyclic and $G$ is not $p$-nilpotent. Chapter 4 discusses the general strategy to tackle generating sets for $TF(G)$. All we know at present when $G$ is $p$-soluble is that we have an exact sequence

$$0 \longrightarrow G/G'S \longrightarrow T(G) \longrightarrow TT(S) \oplus \mathbb{Z}^n \longrightarrow 0 \, ,$$

where $S \in \mathrm{Syl}_p(G)$ and $n$ is the number of connected components of the poset $\mathscr{E}_{\geq 2}(G)/G$ defined in Sect. 3.2. Moreover, $TT(S) = 0$ unless $p = 2$ and $S$ is semi-dihedral (assuming that $G$ has $p$-rank at least 2).

# Chapter 4
# The Torsionfree Part of the Group of Endotrivial Modules

Let $G$ be a finite group and $T(G)$ the group of endotrivial $kG$-modules, where $k$ is a field of characteristic $p$. We have seen in Proposition 3.3 that $T(G) = TT(G) \oplus TF(G)$ is finitely generated, with $TT(G)$ the torsion subgroup of $T(G)$ and $TF(G)$ a torsionfree abelian group of finite rank, equal to the number of connected components of $\mathcal{E}_{\geq 2}(G)/G$, the poset of $G$-conjugacy classes of elementary abelian $p$-subgroups of $G$ of rank at least 2.

In this chapter, we present various methods which aim at finding a suitable set of generators for $TF(G)$ whenever $\mathcal{E}_{\geq 2}(G)/G$ has at least 2 connected components. Recall that if $\mathcal{E}_{\geq 2}(G)/G$ is empty (i.e. $G$ has $p$-rank 1), then $TF(G)$ is trivial and so $T(G)$ is finite, and if $\mathcal{E}_{\geq 2}(G)/G$ is connected, then $TF(G) = \langle \Omega_G \rangle \cong \mathbb{Z}$.

Moreover, Theorem 3.5 shows that there are maximal bounds on the $p$-rank of $G$ beyond which $TF(G)$ must be cyclic, and on the rank of $TF(G)$. Namely:

- If $G$ has $p$-rank greater than $p$ if $p$ is odd, or greater than 4 if $p = 2$, then $\mathcal{E}_{\geq 2}(G)/G$ is connected and therefore $TF(G) = \langle \Omega_G \rangle \cong \mathbb{Z}$.
- The poset $\mathcal{E}_{\geq 2}(G)/G$ has at most $p + 1$ connected components if $p$ is odd, and at most 5 if $p = 2$, from which we gather that $TF(G)$ has $\mathbb{Z}$-rank at most $p + 1$, respectively 5.

But why is it that the $\mathbb{Z}$-rank of $TF(G)$ is equal to the number of connected components of $\mathcal{E}_{\geq 2}(G)/G$?

## 4.1 The Torsionfree Rank of $T(G)$: From $p$-Groups to Arbitrary Groups

Alperin determined the rank of $TF(G)$ as a torsionfree abelian group in [3, Theorem 4] in the case of a finite $p$-group $G$. The result was then extended by Carlson, Mazza and

© The Author(s), under exclusive license to Springer Nature Switzerland AG 2019
N. Mazza, *Endotrivial Modules*, SpringerBriefs in Mathematics,
https://doi.org/10.1007/978-3-030-18156-7_4

Nakano to arbitrary finite groups [34, Sect. 3]. Alperin's proof uses relative syzygies and the following facts (cf. [3, Sect. 5]).

**Lemma 4.1.** *Let $G$ be a finite $p$-group and $F$ a maximal elementary abelian subgroup of $G$ of order $p^2$.*

1. *If $F$ is contained in the Frattini subgroup of $G$, then:*

    a.  *$G$ has rank 2.*
    b.  *$F$ is normal in $G$.*
    c.  *$F$ is the unique maximal elementary abelian subgroup of $G$ contained in the Frattini subgroup.*

2. *Let $X, Y$ be finite $G$-sets. There is an exact sequence of $kG$-modules*

$$0 \longrightarrow \Omega_X(k) \otimes \Omega_Y(k) \longrightarrow k[X \times Y] \longrightarrow \Omega_{X \cup Y}(k) \longrightarrow 0 .$$

From this lemma, and using Puig's result in [107], which asserts that the kernel of the restriction map

$$\text{Res} : T(G) \to \prod_{[E] \in \mathscr{E}_{\geq 2}(G)/G} T(E) \cong \mathbb{Z}^n \text{ is finite,} \tag{4.1}$$

where $E \in \mathscr{E}_{\geq 2}(G)/G$ (cf. Theorem 3.3), Alperin deduces the following.

**Proposition 4.1.** *Let $G$ be a finite $p$-group. There is an endotrivial $kG$-module whose class in $T(G)$ has a nontrivial restriction to $F$ and a trivial restriction to every other maximal elementary abelian $p$-subgroup of $G$ not conjugate with $F$.*

The conclusion is that since the map

$$\text{Res} : TF(G) \to \prod_{E \in \mathscr{E}_{\geq 2}(G)/G} T(E) \cong \mathbb{Z}^n \text{ is injective,}$$

where $n$ is the number of connected components of $\mathscr{E}_{\geq 2}(G)/G$, and since for each component $\mathscr{C}$ of $\mathscr{E}_{\geq 2}(G)/G$ there exists an endotrivial $kG$-module $M$ such that $M{\downarrow}_E^G \not\cong k \oplus (\text{proj})$ for every $E \in \mathscr{C}$, and $M{\downarrow}_E^G \cong k \oplus (\text{proj})$ for every $E \notin \mathscr{C}$, we conclude that $TF(G)$ must be free abelian of rank $n$.

The corresponding result for arbitrary finite groups relies on cohomological methods and the fact that the torsion subgroup $TT(G)$ is equal to the kernel of the restriction map in Eq. (4.1). So the rank of $TF(G)$ is at most the number $n$ of connected components of $\mathscr{E}_{\geq 2}(G)/G$, and the difficulty is to show that it is equal to $n$, that is, $\text{im}(\text{Res})$ is a subgroup of finite index in $\prod_{[E] \in \mathscr{E}_{\geq 2}(G)/G} T(E)$.

Let $S \in \text{Syl}_p(G)$. Without loss, suppose that $G$ has $p$-rank at least 2, and that $S < G$. We can also suppose that $n > 1$. Indeed, if $n = 0$, then $TF(G) = \{0\}$, and if $n = 1$, then $TF(G) = \langle \Omega_G \rangle$.

To prove that $TF(G)$ has rank $n$, we invoke Quillen's theorem, already used in Sects. 3.4 and 3.10. Let $V_G(k)$ denote the maximal ideal spectrum of the cohomology ring $H^*(G, k)$. Then,

$$V_G(k) = \bigcup_E \text{res}_{G,E}^*(V_E(k)),$$

where $E$ runs through all the elementary abelian $p$-subgroups of $G$.

If $n > 1$, then $S$ has a cyclic centre and so contains a unique central subgroup $Z$ of order $p$. If $E$ is an elementary abelian $p$-subgroup of $G$, then $V_E(k)$ is a plane, and if $E$, $F$ belong to two distinct connected components of $\mathscr{E}_{\geq 2}(G)/G$, then

$$\text{res}_{G,E}^*(V_E(k)) \cap \text{res}_{G,F}^*(V_F(k)) = \text{res}_{G,Z}^*(V_Z(k))$$

is a 1-dimensional subvariety of $V_G(k)$. Therefore, for some positive integer $m$, there exists an element $\zeta \in H^m(G, k)$ such that $\text{Res}_Z^G(\zeta) \in H^m(Z, k)$ is not nilpotent, i.e. $\zeta \neq 0$, and if $p$ is odd, then $m$ must be even. Let $\tilde{\zeta} \in \text{Hom}_{kG}(\Omega_G^m(k), k)$ be a cocycle representing $\zeta$ and $L_\zeta = \ker(\tilde{\zeta})$. These are the same $L_\zeta$ modules used in Sect. 3.4. We know that $V_G(L_\zeta)$ is the disjoint union of subvarieties:

$$V_G(L_\zeta) = V_1 \sqcup \cdots \sqcup V_n \text{ where } V_i = \bigcup_{E \in \mathscr{C}_i} \text{res}_{G,E}^* \left( V_E(\text{Res}_E^G(\zeta)) \right),$$

where $\mathscr{C}_1, \ldots, \mathscr{C}_n$ are the distinct connected components of $\mathscr{E}_{\geq 2}(G)/G$. Accordingly, the $kG$-module $L_\zeta$ decomposes as a direct sum $L_\zeta = L_1 \oplus \cdots \oplus L_n$ with $V_i = V_G(L_i)$ for each $1 \leq i \leq n$.

Consider the push-out construction as in Sect. 3.4:

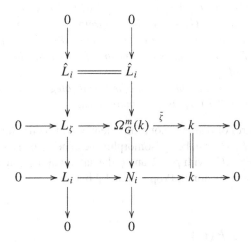

where $\hat{L}_i$ is the direct sum of all the summands of $L_\zeta$ except $L_i$. As in Sect. 3.4, each $kG$-module $N_i$ has the property that $N_i \downarrow_E^G \cong \Omega_E^m(k) \oplus$ (proj) for any $E \in \mathscr{C}_i$, and $N_i \downarrow_E^G \cong k \oplus$ (proj) for any $E \notin \mathscr{C}_i$. Taking the stable isomorphism classes of the modules $N_i$, we obtain $n$ elements of $T(G)$ which generate a torsionfree subgroup of rank precisely $n$. We conclude that the rank of $TF(G)$ is equal to $n$, as asserted (cf. [34, Theorem 3.1]).

**Theorem 4.1.** *Let $G$ be a finite group. The rank of $TF(G)$ is equal to the number of connected components of $\mathscr{E}_{\geq 2}(G)/G$. In other words, it is zero if $G$ has $p$-rank one, it is equal to the number of $G$-conjugacy classes of elementary abelian $p$-subgroups of $G$ of rank 2 if $G$ has $p$-rank 2, or this number plus one if $G$ has $p$-rank at least 3.*

Recall that a subgroup $H$ of $G$ *controls the $p$-fusion* in $G$ if $H$ contains a Sylow $p$-subgroup of $G$ and whenever $Q$, ${}^gQ \leq H$, for some $p$-subgroup $Q$ of $G$, there exists $h \in H$ and $c \in C_G(Q)$ such that $g = hc$ (cf. [118, Lemma 49.1]). More generally, a group homomorphism $\varphi : H \to G$ *controls the $p$-fusion* if $\varphi$ maps a Sylow $p$-subgroup of $H$ to a Sylow $p$-subgroup of $G$, and whenever ${}^hQ \leq R$ for two $p$-subgroups $Q$, $R$ of $H$ and $h \in H$, there exists a $g \in G$ such that ${}^g\varphi(Q) = \varphi({}^hQ) \leq \varphi(R)$ in $G$.

Consequently, Theorem 4.1 says that the $p$-fusion in $G$, or more precisely on $\mathscr{E}_{\geq 2}(G)/G$, matters. In particular, we have [34, Corollary 3.2].

**Corollary 4.1.** *Let $G$ be a finite group and $S \in \mathrm{Syl}_p(G)$.*

1. *If $p = 2$ and $S$ is dihedral, then $TF(G) \cong \mathbb{Z}^2$.*
2. *If $H \leq G$ controls the $p$-fusion in $G$, then $\mathrm{im}\left(\mathrm{Res}_H^G : TF(G) \longrightarrow TF(H)\right)$ is a subgroup of finite index in $T(H)$.*

An unpublished result of Barthel, Grodal and Hunt [9] shows that the control of fusion is not a sufficient condition for the restriction map to be an isomorphism. More precisely, they prove the following result.

**Theorem 4.2.** *Let $G_1$ and $G_2$ be finite groups with Sylow $p$-subgroups $S_1$ and $S_2$ respectively, and let $\varphi : G_1 \to G_2$ be a group homomorphism inducing an isomorphism of the fusion systems $\mathscr{F}_{S_1}(G_1) \cong \mathscr{F}_{S_2}(G_2)$. If $p = 2$, then $\mathrm{Res}_\varphi : TF(G_2) \to TF(G_1)$ is an isomorphism. If $p$ is odd, then $\mathrm{Res}_\varphi$ is not an isomorphism in general.*

*If $\varphi : G_1 \to G_2$ induces an equivalence of the orbit categories $\mathcal{O}_p(G_1) \cong \mathcal{O}_p(G_2)$, then $\mathrm{Res}_\varphi : TF(G_2) \to TF(G_1)$ is an isomorphism.*

The fact that control of fusion is not a sufficient condition for the restriction map $\mathrm{Res}_\varphi : TF(G_2) \to TF(G_1)$ to be an isomorphism can be seen in the example of the groups $\mathrm{SL}_3(7)$ and $\mathrm{PSL}_3(7)$ with $p = 7$ and $\varphi$ the natural quotient map, for instance. We will come back to $p$-fusion in Proposition 4.2 below.

## 4.2 Generating $TF(G)$

In this second part of Chap. 4, we present a partially successful method to find suitable generators for $TF(G)$ when $TF(G)$ is noncyclic. Hence, for the remainder of this chapter, let us assume that $G$ is a finite group with a Sylow subgroup $S$ such that

$\mathcal{E}_{\geq 2}(G)/G$ is nonempty disconnected. Let $E_1, \ldots, E_n \leq S$ be elementary abelian $p$-subgroups of $G$ of rank 2 representing the $n$ distinct connected components of $\mathcal{E}_{\geq 2}(G)/G$. Suppose also that if $G$ has $p$-rank at least 3, then $E_1$ is normal in $S$, and that $E_2, \ldots, E_n$ are maximal elementary abelian subgroups of $S$ not necessarily normal in $S$ (in fact, they are seldom normal in $S$). Let $Z$ be the unique central subgroup of $S$ of order $p$ and write $E_i = Z \times A_i$ for a noncentral subgroup $A_i$ of $E_i$ order $p$. Then $C_S(E_i) = A_i \times Q_i$, where $Q_i$ is cyclic or possibly quaternion if $p = 2$, and $N_S(E_i) = E_1 C_S(E_i)$ with $E_1 E_i \cong p_+^{1+2}$ for all $2 \leq i \leq n$.

When looking for a set of generators of a suitable $TF(G)$, we can assume that $\Omega_G$ belongs to it. Up to this point, we have seen three methods to construct endotrivial modules: one uses *relative syzygies*, another *disassembles some syzygy of the trivial module*, and the third is the *cohomological push-out method*.

From Sect. 4.1, we know that the image of the restriction map

$$\prod_{[E] \in \mathcal{E}_{\geq 2}(G)/G} \mathrm{Res}_{E_i}^G : T(G) \longrightarrow \prod_{[E] \in \mathcal{E}_{\geq 2}(G)/G} T(E) \text{ is a subgroup of finite index.}$$

$$(4.2)$$

Moreover, $T(E) = \langle \Omega_E \rangle \cong \mathbb{Z}$ for each $E \in \mathcal{E}_{\geq 2}(G)/G$.

**Definition 4.1.** The *type* of an endotrivial $kG$-module $M$ is the $n$-tuple $(a_1, \ldots, a_n)$ defined by

$$\mathrm{Res}_{E_i}^G(M) \cong \Omega_{E_i}^{a_i}(k) \oplus (\mathrm{proj}) \text{ for all } 1 \leq i \leq n.$$

In particular, the type of $\Omega_G(k)$ is the constant $n$-tuple $(1, \ldots, 1)$. The fact that $\mathrm{im}(\mathrm{Res})$ has finite index implies that for each $1 \leq i \leq n$ there is some endotrivial $kG$-module $N_i$ such that $\mathrm{Res}_{E_j}^G([N_i]) = a_i \delta_{i,j} \Omega_{E_j}$, where $\delta_{i,j} = 1$ if $i = j$ and $\delta_{i,j} = 0$ otherwise, and where $a_i \neq 0$. But we do not know the minimal positive value of $a_i$. In other words, in order to find a set of generators for $TF(G)$, we want to find endotrivial $kG$-modules $N_2, \ldots, N_n$ such that $\mathrm{Res}_{E_j}^G([N_i]) = a_j \delta_{i,j} \Omega_{E_j}$ with $a_j > 0$ minimal for all $2 \leq i, j \leq n$. Adding to such set the class $\Omega_G$, we then conclude that $T(G) = TT(G) \oplus \langle \Omega_G, [N_2], \ldots, [N_n] \rangle$.

Let $S \in \mathrm{Syl}_p(G)$. Observe that $\mathrm{Res}_S^G : TF(G) \to TF(S)$ is injective (cf. [40, Proposition 2.3]). Then, using the known structure of $T(S)$ (since $S$ is a finite $p$-group), we obtain lower bounds on the $a_i$'s. Explicitly, [44, Theorem 7.1] shows that, if $t$ is the number of connected components of $\mathcal{E}_{\geq 2}(S)/S$, then there exist endotrivial $kS$-modules $N_2, \ldots, N_t$ of type $(\delta_{i,j} b_i p)_{j=1}^t$, for $2 \leq i \leq t$, where

$$b_i = \begin{cases} 1 \text{ if } Q_i/A_i \text{ is cyclic of order 2,} \\ 2 \text{ if } Q_i/A_i \text{ is cyclic of order } \geq 3, \\ 4 \text{ if } Q_i/A_i \text{ is quaternion.} \end{cases} \qquad (4.3)$$

and furthermore we know that $b_2 = \cdots = b_n$ if $p$ is odd or if $b_i = 1$ for some $i$.

In Sect. 3.4, we described a method that uses relative syzygies to construct a set of generators for $TF(G)$ whenever $S$ is normal in $G$. But we do not know if and how we could extend this method any further. Note also that the resulting modules (obtained by tensor induction) are not indecomposable in general.

In the second method described in Sect. 3.4, we constructed an endotrivial module as a subquotient of some syzygy $\Omega_G^m(k)$ of the trivial module. This method can be generalised to $N_G(Z)$ for any finite group $G$ [40, Theorem 5.2].

**Theorem 4.3.** *Assume the above notation. For each $2 \leq i \leq n$, there exist endotrivial $kN_G(Z)$-modules $N_2, \ldots, N_n$, where each $N_i$ is obtained as a subquotient of $\Omega_G^{b_i p}(k)$ and has type $(\delta_{i,j} b_i p)_{j=1}^n$, for $b_i$ defined in Eq. (4.3). Hence*

$$T(N_G(Z)) = TT(N_G(Z)) \oplus \langle \Omega_{N_G(Z)}, [N_2], \ldots, [N_n] \rangle.$$

This method seems "better" than the previous one, as it will produce smaller modules than those obtained by taking tensor products of endotrivial modules. But still, this "disassembly" method does not give a general solution to the problem of finding generators for $TF(G)$.

The third method is the *cohomological push-out*, used in Sects. 3.4 and 4.1. In this latter section, we have obtained endotrivial $kG$-modules $N_i$ constructed as a push-out using some non-nilpotent cohomology class, for an arbitrary finite group $G$. But the drawback is that we are not guaranteed an integral basis for $TF(G)$ as we do not know if the degree of the cohomology class picked is the minimal possible value for the type. The argument generalises to the case when we replace the trivial module by any 1-dimensional $kG$-module. We call the cohomological push-out method *optimal* when the endotrivial $kG$-modules produced as push-outs form an integral basis of $TF(G)$.

In particular, if we know that we can choose the cohomology class in degree $b_i p$, where $b_i$ is defined in Eq. (4.3), then the method must be optimal. However, this is rarely the case in odd characteristic using the trivial module. Indeed, we then have $b_i p = 2p$ for all $i$, and if $N_G(Z) \neq C_G(Z)$, then $H^{2p}(G, k)$ does not contain any suitable cohomology class $\zeta$ [40, Lemma 6.2].

We end this chapter with a result giving an instance of when the cohomological push-out method is optimal (cf. [40, Proposition 10.4 and Corollary 10.5]).

**Proposition 4.2.** *Let $\varphi : H \to G$ be a group homomorphism which controls $p$-fusion. Suppose that $k$ is algebraically closed and that the cohomological push-out method is optimal for $H$. Then the induced homomorphism $\mathrm{Res}_\varphi : TF(G) \longrightarrow TF(H)$ is an isomorphism. Moreover, the cohomological pushout method is optimal for $G$. In particular, if the normaliser $H$ of a Sylow $p$-subgroup of $G$ controls $p$-fusion in $G$ and the cohomological push-out method is optimal for $H$, then the restriction map $\mathrm{Res}_H^G : TF(G) \longrightarrow TF(H)$ is an isomorphism and the cohomological push-out method is optimal for $G$.*

# Chapter 5
# Torsion Endotrivial Modules

In this chapter we concentrate our attention on the torsion subgroup $TT(G)$ of the group of endotrivial $kG$-modules for a finite group $G$. We have seen that if $S$ is a finite $p$-group, then $TT(S) = 0$ unless $S$ is cyclic, semi-dihedral or quaternion. Furthermore, whenever $TT(S) \neq 0$, Sects. 3.6, 3.9 and 3.10 show that these torsion endotrivial $kS$-modules give torsion endotrivial modules for $G$ whenever $S \in \mathrm{Syl}_p(G)$. Thus, the remaining question is about the subgroup

$$K(G) = \ker \left( \mathrm{Res}_S^G : T(G) \to T(S) \right)$$

of $T(G)$, where $S \in \mathrm{Syl}_p(G)$, generated by the stable isomorphism classes of the indecomposable trivial source endotrivial $kG$-modules, i.e. the *trivial Sylow restriction $kG$-modules* (cf. Definition 3.4). In general, we expect $K(G)$ to be isomorphic to the group $\mathrm{Hom}(G, k^\times) \cong G/G'S$ of 1-dimensional $kG$-modules, unless $G$ has a strongly $p$-embedded subgroup. In particular, if $S \cap {}^gS \neq 1$ for all $g \in G$, then $K(G) \cong G/G'S$ by Lemma 3.2.

Since 2010, the methods to investigate $K(G)$ have proliferated and diversified: (ordinary) character theory, weak homomorphisms, and homotopy theory in particular. But none can provide a straighforward description of $K(G)$ in full generality. So, our approach consists in trying to find $K(G)$ for specific classes of groups, or find the subgroup of $K(G)$ spanned by the simple trivial Sylow restriction modules. As mentioned in Sect. 2.5, G. Robinson proves in [110] that the study of simple endotrivial $kG$-modules that are not monomial may be reduced to the case when $G$ is quasi-simple.

It is worth pointing out that we do not have any example of a finite group $G$ of $p$-rank greater than 2 and no strongly $p$-embedded subgroup which has an indecomposable trivial Sylow restriction module of dimension greater than 1.

Our objective in this section is to describe these techniques and present some of their applications. Let us first summarise the results on $K(G)$ obtained using "traditional" methods in modular representation theory (cf. [37]), most of which we have seen in Sect. 3.5.

© The Author(s), under exclusive license to Springer Nature Switzerland AG 2019
N. Mazza, *Endotrivial Modules*, SpringerBriefs in Mathematics,
https://doi.org/10.1007/978-3-030-18156-7_5

**Theorem 5.1.** *Let $G$ be a finite group of $p$-rank at least 2, and let $S \in \mathrm{Syl}_p(G)$.*

1. *If $S \cap {}^g S$ has order divisible by $p$ for all $g \in G$, then $K(G) \cong G/G'S$. This holds in particular if $O_p(G) > 1$.*
2. *If $H$ is strongly $p$-embedded in $G$, then $\mathrm{Res}_H^G : K(G) \to K(H)$ is an isomorphism. This holds in particular if $S$ is TI and $H = N_G(S)$.*
3. *If $G = H *_Z K$ is the central product of two groups of order divisible by $p$, then $K(G) \cong G/G'S$.*
4. *Suppose that $H$ is a normal subgroup of $G$ of index not divisible by $p$, and such that $K(H) \cong H/H'S$. Then $K(G) \cong G/G'S$.*

*Proof.*  1. This is proved in Lemma 3.2, Part 1.

2. This is stated in Proposition 3.3, Part 3. The proof is similar to that for Part 1. Namely, an indecomposable trivial Sylow restriction $kH$-module $V$ is a direct summand of $k{\uparrow}_S^H$ and its $kG$-Green correspondent $\Gamma^G(V)$ is an indecomposable summand of $V{\uparrow}_H^G$, which itself is a direct summand of $k{\uparrow}_S^G$. So $\Gamma^G(V){\downarrow}_S^G$ is a direct summand of

$$(V{\uparrow}_H^G){\downarrow}_S^G \mid (k{\uparrow}_S^G){\downarrow}_S^G \cong k^{|N_G(S):S|} \oplus \left( \bigoplus_x k{\uparrow}_{S \cap {}^x S}^S \right),$$

where $x$ runs through a set of double cosets representatives of $[S\backslash G/S]$ which do not normalise $S$. On the right-hand side of the isomorphism symbol, the module $k$ must appear exactly once in $\Gamma^G(V)$, because $H \geq N_G(S)$. Moreover, $S \cap {}^x S = 1$ for all $x \notin H$, since $H$ is strongly $p$-embedded in $G$, and so $k{\uparrow}_{S \cap {}^x S}^S \cong kS$ is projective for all $x \in [S\backslash G/S] - H$. Therefore $\Gamma^G(V)$ is endotrivial and the claim is proved.

3. The proof is essentially done in [37, Theorem 2.4]. Write $G = H *_Z K$ where $Z \leq Z(H) \cap Z(K) = H \cap K$. For this last equality, note that since $H \cap K \leq K$ and $K \leq C_G(H)$, the subgroup $H \cap K$ centralises $H$, and similarly $H \cap K$ also centralises $K$. That is, $H \cap K \leq Z(G)$ (in particular, it is abelian). The assertion is clear if $p$ divides $|H \cap K|$. Suppose that $H \cap K$ is a $p'$-group and let $M$ be an indecomposable trivial Sylow restriction $kG$-module. Let $S$ be a Sylow $p$-subgroup of $G$, and let $S_K = S \cap K$. Since $K$ is normal in $G$, we have $S_K \in \mathrm{Syl}_p(K)$ and $S_K \nleq H$ (otherwise, $S_K \leq K \cap H \leq Z(G)$). Let $J = HS_K$. Then $O_p(J) \geq S_K > 1$, so that $M{\downarrow}_J^G \cong V \oplus (\mathrm{proj})$, with $\dim(V) = 1$. Now, $M$ is a direct summand of $V{\uparrow}_J^G$ because $M$ is relatively $J$-projective. We claim that $S_H = S \cap H \in \mathrm{Syl}_p(H)$ acts trivially on $V{\uparrow}_J^G$. Indeed, by Mackey's formula

$$(V{\uparrow}_J^G){\downarrow}_{S_H}^G \cong \sum_{x \in [S_H \backslash G/J]} ({}^x V{\downarrow}_{{}^x J \cap S_H}^J){\uparrow}_{{}^x J \cap S_H}^{S_H}, \tag{5.1}$$

where the set $[S_H \backslash G/J]$ can be chosen in $K$. Since $K$ centralises $S_H$, we may assume that ${}^x J \cap S_H = {}^x(J \cap S_H) = {}^x S_H = S_H$ for all $x \in [S_H \backslash G/J]$. So the right-hand side of Eq. (5.1) is a direct sum of trivial modules since $\dim(V) = 1$.

In particular, it does not contain any nonzero projective summand, and since $M$ is endotrivial, $M$ must have dimension 1.

4. This is proved in Lemma 3.2, Part 2                                                    □

## 5.1  $K(G)$ Using Ordinary Character Theory

Theorem 3.2 states that every endotrivial $kG$-module lifts to an endotrivial $\mathcal{O}G$-module and since $\mathcal{O}$ has characteristic zero, we can use ordinary character theory. The efficiency of this approach can be seen for instance in [111], which preceeds [88, 89].

The group $K(G)$ of trivial Sylow restriction $kG$-modules is generated by the stable isomorphism classes of trivial source endotrivial $kG$-modules, i.e. indecomposable trivial Sylow restriction $kG$-modules. In particular, we know that the trivial source $kG$-modules are precisely the indecomposable $p$-permutation $kG$-modules and that they lift to $\mathcal{O}G$-modules uniquely (cf. [118, Proposition 27.11]). Let us recall some well-known useful facts about $p$-permutation modules (cf. [85, Lemma II.12.6]).

**Lemma 5.1.** *Let $M$ be a trivial source $kG$-module and $\hat{M}$ its lift to $\mathcal{O}G$ with character $\chi$. For any $p$-element $x \in G$, then $\chi(x)$ is a non-negative integer, equal to the number of indecomposable summands of $M\downarrow^{G}_{\langle x \rangle}$ that are isomorphic to the trivial module.*

*Moreover, $\chi(x) > 0$ if and only if $x$ belongs to a vertex of $M$ (cf. Sect. 1.4.3).*

From this criterion on character values of $p$-elements, [88, Theorem 2.2 and Corollary 2.3] and [89, Corollary 2.3] prove the following.

**Theorem 5.2.** *Let $G$ be a finite group.*

1. *Let $M$ be a trivial source $kG$-module which lifts to an $\mathcal{O}G$-module with character $\chi$. Then $M$ is endotrivial if and only if $\chi(x) = 1$ for each $p$-element $1 \neq x \in G$.*
2. *Suppose that $M$ is an endotrivial $kG$-module which lifts to a $\mathbb{C}G$-module with character $\chi$. Then $|\chi(x)| = 1$ for every $p$-singular element $x \in G$ (where $|z|$ is the modulus of a complex number $z$).*
3. *If $S \in \mathrm{Syl}_p(G)$ has normal $p$-rank at least 2 and $M$ is a selfdual endotrivial $kG$-module which lifts to an $\mathcal{O}G$-module with character $\chi$, then $\chi(1) \equiv 1 \pmod{|S|}$.*

Recall that an element is *p-singular* if it has order divisible by $p$.

The numerical characterisations of endotrivial modules allow the authors in [88, 89] to inspect which simple modules for quasi-simple groups are endotrivial using the character tables in the Atlas of finite groups [49] and in the GAP Character Table Library [62].

Let $S \in \mathrm{Syl}_p(G)$ and $N = N_G(S)$. Observe that a trivial source endotrivial $kG$-module must have a 1-dimensional $kN$-Green correspondent, and it must lie in a block of $kG$ with defect group $S$. So, assuming the character table of $N$ and $G$ are known, the method is as follows.

1. Pick the character $\chi$ of a 1-dimensional $kN$-module.
2. Decompose $\mathrm{Ind}_N^G \chi$ into irreducible characters (using some algebra software, e.g. GAP [62]).
3. Left multiply by the idempotent of a block with full defect to identify the possible constituents of $\mathrm{Ind}_N^G \chi$. If $G$ is quasi-simple, then it often happens that only the principal block has full defect.
4. Look at all the possibilities to attribute sums of subsets of the obtained character values for the constituents of $\mathrm{Ind}_N^G \chi$.
5. The method is "successful", i.e we obtain a new trivial Sylow restriction module whenever we find a unique possibility satisfying one of the criteria for endotrivial modules in Theorem 5.2.

Let us take two examples from [90] showing different possibilities.

*Example 5.1.*   1. Let $G = J_2$ be one of Janko's sporadic simple groups, and let $p = 3$. Then $S$ is extraspecial of order 27 and exponent 3, and $N \cong S \rtimes C_8$. So $N/N'S \cong C_8$. We know that only the principal block $B_0$ of $G$ has full defect. Let us pick $\chi$ the character of a 1-dimensional $kN$-module with order 4. With GAP, we calculate

$$e_0 \, \mathrm{Ind}_N^G \chi = \chi_{12} + \chi_{13} + 2\chi_{21},$$

the decomposition of the part of the induced character which lies in $B_0$. Here, the subscripts indicate the label in the Atlas. Either a direct inspection of the character table of $J_2$ in the Atlas [49] or considering the character degree of $e_0 \, \mathrm{Ind}_N^G \chi$ shows that it cannot be endotrivial. Indeed, there is no sum of a subset of character degrees of $e_0(\mathrm{Ind}_N^G \chi)(1)$ which is congruent to 1 (mod $|S|$).

2. Let $G = HS$ be the Higman–Sims sporadic simple group, and let $p = 5$. Then $S$ is extraspecial of order $5^3$ and exponent 5, and $N \cong 5_+^{1+2} \rtimes (C_8.C_2)$. We calculate $K(N) \cong N/N'S \cong \mathbb{Z}/2 \oplus \mathbb{Z}/4$. We know that $G$ contains a maximal subgroup isomorphic to $\mathrm{PSU}_3(5).C_2$ which contains $N$ as a strongly 5-embedded subgroup, and so $K(H) \cong K(N) \cong \mathbb{Z}/2 \oplus \mathbb{Z}/4$. Let $\chi$ be the nontrivial character of $H$ of degree 1 corresponding to the $\mathbb{Z}/2$ factor, and let $e_0$ be the idempotent of the principal block $B_0$ of $G$, which is in this case the unique 5-block with full defect. Using GAP, we find $e_0 \cdot (\chi{\uparrow}_H^G) = \chi_2 + \chi_5$, where $\chi_2(1)$ and $\chi_5(1)$ have degrees 22 and 154 respectively. So this does not give any endotrivial module, and we gather that $K(G) \leq \mathbb{Z}/4$. Take now a linear character $\mu$ of $N$ of order 4. We calculate

$$e_0 \cdot (\mu{\uparrow}_N^G) = \chi_8 + \chi_9 + \chi_{10} + \chi_{16} + \chi_{17} + 2 \cdot \chi_{22}.$$

Using Lemma 5.1 and the decomposition matrix of $G$ for the prime 5, we conclude that the $kG$-Green correspondent $\Gamma^G(\mu)$ of $\mu$ affords either $\chi_8 + \chi_{10}$ (of degree $231 + 770 = 1001 \equiv 1 \pmod{125}$) or $\chi_8 + \chi_{22}$ (of degree $231 + 2520 = 2751 \equiv 1 \pmod{125}$). In either case we conclude that $\Gamma^G(\mu)$ is endotrivial by Theorem 5.2. Therefore $TT(G) = K(G) = \langle[\Gamma^G(\mu)]\rangle \cong \mathbb{Z}/4$.

## 5.2 $K(G)$ Using Balmer's Weak Homomorphisms

In [8], P. Balmer introduced an original way to look at trivial Sylow restriction $kG$-modules drawing on algebraic-geometry methods. Given $H \leq G$ of index coprime to $p$, this method describes the kernel of the restriction map

$$\ker \left( \operatorname{Res}_H^G : T(G) \longrightarrow T(H) \right),$$

using the conjugation action of $G$ on the poset of its $p$-subgroups. Hence, Balmer comes up with the following concept (cf. [8, Definition 2.2]).

**Definition 5.1.** Let $H$ be a subgroup of $G$ of index invertible in the field $k$. A *weak $H$-homomorphism* is a function

$$u : G \longrightarrow k^\times \quad \text{subject to the following three properties:}$$

(WH1) For every $h \in H$, we have $u(h) = 1$.
(WH2) For every $g \in G$ such that $|H \cap H^g|$ is coprime to $p$, we have $u(g) = 1$.
(WH3) For every $g_1, g_2 \in G$ such that $|H \cap H^{g_1} \cap H^{g_2 g_1}|$ is divisible by $p$, we have
$$u(g_2 g_1) = u(g_2) u(g_1).$$

We define $A(G, H)$ to be the abelian group of weak $H$-homomorphisms $G \longrightarrow k^\times$ under element-wise multiplication: $(uv)(g) = u(g)v(g)$ for every $g \in G$ and all weak $H$-homomorphisms $u, v$.

In particular, if $H$ is normal in $G$, then (WH2) is never fulfilled, whilst (WH3) always holds, implying that $u$ is a group homomorphism whose kernel contains $H$. It then follows that $A(G, H) = \operatorname{Hom}(G/H, k^\times)$. By contrast, if $H$ is strongly $p$-embedded in $G$, then (WH2) holds for all $g \in G - H$, whilst (WH3) never holds whenever one of $g_1$ or $g_2$ does not lie in $H$, implying that the trivial homomorphism $u(g) = 1$ for all $g \in G$ is the only weak $H$-homomorphism. Thus $A(G, H) = 1$ in this case.

Balmer's striking result using weak $H$-homomorphisms is [8, Theorem 2.8], which he proves by explicitly constructing such a group isomorphism.

**Theorem 5.3.** *Assume the above notation. There is a group isomorphism*

$$\ker \left( \operatorname{Res}_H^G : T(G) \to T(H) \right) \longrightarrow A(G, H).$$

*In other words, $K(G) \cong A(G, S)$.*

**Proposition 5.1.** *Let $G$ be a finite group such that $O_p(G) > 1$, and let $S \in \operatorname{Syl}_p(G)$.*

1. *Every weak $S$-homomorphism is a group homomorphism.*
2. *The isomorphism $K(G) \to A(G, S)$ maps the class of a one-dimensional $kG$-module to the corresponding group representation $G \to k^\times$, and there is a group isomorphism $A(G, S) \cong \operatorname{Hom}(G, k^\times)$.*

Proposition 5.1 is proved in [45, Proposition 3.3] using Theorem 5.3. Extending Balmer's results, Carlson and Thévenaz obtain a sharper description of $K(G)$, and a very efficient algorithm to use whenever a Sylow $p$-subgroup of $G$ is abelian. They define a chain of subgroups of a $p$-local subgroup of $G$ in [45, Sect. 4].

**Definition 5.2.** Fix a Sylow $p$-subgroup $S$ of $G$. For each $1 < Q \leq S$, let $N_Q = N_G(Q)$ and $S_Q \in \mathrm{Syl}_p(N_Q)$, and define

$$\rho^1(Q) = N'_Q S_Q,$$
$$\rho^i(Q) = \langle N_Q \cap \rho^{i-1}(R) \,|\, \{1\} \neq R \leq S \rangle \quad \text{for} \ \ i > 1,$$
$$\rho^\infty(Q) = \bigcup_{i \geq 1} \rho^i(Q).$$

Since $G$ is finite, there must be some positive integer $j$ such that the chain stabilises, i.e. $\rho^1(Q) \leq \rho^2(Q) \leq \cdots \leq \rho^j(Q) = \rho^\infty(Q) \leq N_Q$. Moreover, $K(N_Q) \cong N_Q/\rho^1(Q)$.

The point of defining these subgroups is that $\rho^\infty(S)$ is contained in the kernel of any 1-dimensional $kN_G(S)$-module whose $kG$-Green correspondent is endotrivial [45, Proposition 4.2]. As a consequence, if $\rho^\infty(S) = N_G(S)$, or more generally if $\rho^i(Q) = N_G(Q)$ for some $i \geq 1$ and a nontrivial characteristic subgroup $Q$ of $S$, then $K(G) = \{0\}$, that is, $k$ is the only indecomposable trivial Sylow restriction $kG$-module, and this also forces $G'S = G$.

In the particular case when $S$ is abelian, we know by Burnside's fusion theorem that $N_G(S)$ controls the $p$-fusion in $G$. Using this fact, Carlson and Thévenaz show that $\rho^\infty(S) = \rho^2(S)$, and so $K(G) \cong N_G(S)/\rho^2(S)$ [45, Theorem 5.1]. They used their result to determine $K(G)$ for some examples of classical and sporadic simple groups with abelian Sylow $p$-subgroups, and they conjectured that in general $K(G) \cong N_G(S)/\rho^\infty(S)$ (cf. [45, Conjecture 5.5]). This conjecture was later proved by Grodal, and we will come back to this in Sect. 5.3. For now, let us carry on with the $\rho$-subgroup series, and how it was adapted in [37, Theorem 3.1].

**Theorem 5.4.** Let $S \in \mathrm{Syl}_p(G)$ and $H \leq G$ with $N_G(S) \leq H$. Suppose that the following conditions hold.

(A) $K(H) \cong H/H'S$, i.e. every indecomposable trivial Sylow restriction $kH$-module has dimension 1.

(B) $H = \langle g_1, \ldots, g_m \rangle$ such that for each $i$, either

    (1) $g_i \in H'S$, or

    (2) there exists a subgroup $H_i$ of $G$ such that the following three conditions hold.

        (a) $K(H_i) \cong H_i/H'_i S_i$, where $S_i \in \mathrm{Syl}_p(H_i)$.

        (b) $p \mid |H_i \cap H|$.

        (c) $g_i \in H'_i S_i$.

*Then $K(G) = \{0\}$, that is, $k$ is the unique indecomposable trivial Sylow restriction $kG$-module.*

*Proof.* Suppose that $M$ is an indecomposable trivial Sylow restriction $kG$-module. From (A), we know that $M{\downarrow}_H^G \cong V \oplus (\text{proj})$, where $\dim(V) = 1$. So $H'S \leq \ker(V)$, and any generator $g_i$ of $H$ that lies in $H'S$ is in $\ker(V)$. We are left with the generators $g_i$ of $H$ satisfying (2), and we need to show that they too are in $\ker(V)$. Pick such a $g_i$, and let $H_i \leq G$ satisfying the conditions (a), (b) and (c) in (2). Let $Q_i$ be a nontrivial $p$-subgroup of $H_i \cap H$, whose existence is ensured by (2)(b). The condition (2)(a) implies that $M{\downarrow}_{H_i}^G \cong W \oplus (\text{proj})$, where $\dim(W) = 1$. Since $g_i \in H_i'S_i$ by (2)(c), $g_i \in \ker(W)$. Observe the isomorphisms:

$$M{\downarrow}_{H_i \cap H}^G \cong V{\downarrow}_{H_i \cap H}^H \oplus (\text{proj}) \cong W{\downarrow}_{H_i \cap H}^{H_i} \oplus (\text{proj}) .$$

Since $p \mid |H_i \cap H|$, we must have $p \mid \dim (\text{proj})$. So the Krull–Schmidt theorem (cf. Sect. 1.3.3) shows that $W{\downarrow}_{H_i \cap H}^{H_i} \cong V{\downarrow}_{H_i \cap H}^{H}$, both modules having dimension 1. It follows that $g_i \in \ker(V)$.

The upshot is that we have proved that each element $g_i$ in (B) is in the kernel of $V$, i.e. acts trivially on $V$. In other words, we must have $V = k$. Since $M$ is indecomposable with vertex a Sylow $p$-subgroup $S$ of $G$ and $M{\downarrow}_H^G \cong V \oplus (\text{proj})$, where $H \geq N_G(S)$, then $M$ is the $kG$-Green correspondent of $V$, i.e. $M = k$.  $\square$

This last theorem has been successful in determining $K(G)$ in the case of finite groups of Lie type $A$ in non-definining characteristic, i.e. central subquotients of $GL_n(q)$ when $(p, q) = 1$ (cf. [37]).

## 5.3  *K(G)* Using Grodal's Homotopy Method

Grodal's homotopy method consists in associating topological objects to abstract groups. We use the elements of algebraic topology introduced in Sect. 1.6.

Throughout, $G$ is a finite group of order divisible by $p$, the characteristic of the field $k$, and $S \in \mathrm{Syl}_p(G)$. Let $\mathscr{S}_p$ or $\mathscr{S}_p(G)$ denote the poset of all the $p$-subgroups of $G$.

We start with a brief paragraph on (co-)homology computations which explains the notation and conventions used in [70].

Given a group $G$, it is well known that $\mathrm{H}_1(G) \cong G/G'$ is the abelianisation of $G$ (cf. [25, Sect. II.3]), where the coefficients in (co-)homology and homotopy groups are to be understood to be $\mathbb{Z}$ unless otherwise specified (i.e. $\mathrm{H}_*(G) = \mathrm{H}_*(G; \mathbb{Z})$). So for a finite group and coefficients in $k^\times$ instead of $\mathbb{Z}$, we have $\mathrm{H}_1(G, k^\times) \cong G/G'S$ is the $p'$-part of the abelianisation of $G$ because there is no $p$-torsion in $k^\times$.

Grodal considers (co-)homology groups of *small categories*. In particular, regarding $k^\times$ as a category with a single object, we have identifications

$$H^1(\mathcal{O}_p^*(G); k^\times) = \text{Rep}(\mathcal{O}_p^*(G); k^\times) \cong \text{Hom}(H_1(\mathcal{O}_p^*(G)), k^\times) \qquad (5.2)$$

where $\text{Rep}(\mathcal{O}_p^*(G); k^\times)$ denotes the isomorphism classes of functors $\mathcal{O}_p^*(G) \to k^\times$.

[70, Theorems A to I] provide nine different descriptions of $K(G)$ using this approach. Here is [70, Theorem A].

**Theorem 5.5.** *There exists a group isomorphism*

$$\Phi : K(G) \longrightarrow H^1(\mathcal{O}_p^*(G); k^\times), \quad \text{with} \quad \Phi(M) = \varphi_M,$$

*where, for a trivial Sylow restriction $kG$-module $M$, the function $\varphi_M : \mathcal{O}_p^*(G) \to k^\times$ maps $G/Q \in \mathcal{O}_p^*(G)$ to the 1-dimensional $kG$-module $M^Q/u_Q M$, where $u_Q = \sum\limits_{u \in Q} u$ for $Q \in \mathcal{S}_p$.*

*The inverse of $\Phi$ is given by the twisted Steinberg complex,*

$$\Phi^{-1}(\varphi) = C_*(|\mathcal{S}_p^*(G)|; k_\varphi), \quad \text{where} \quad \varphi : \mathcal{O}_p^*(G) \to k^\times, \quad \text{and}$$

*$k_\varphi$ is the $G$-twisted coefficient system induced by $\varphi$.*

The *Steinberg complex of $G$ at the prime $p$* is the chain complex

$$\text{St}_p(G) = C_*(|\mathcal{S}_p^*(G)|; \mathbb{Z}_p),$$

where $\mathbb{Z}_p$ denote the ring of $p$-adic integers (cf. [69, Sect. 5] and [123, Sect. 5]). Webb showed that $\text{St}_p(G)$ splits as the direct sum of a complex of projective modules and a $\mathbb{Z}_p G$-split acyclic complex, and this fact does not change if we replace $\mathcal{S}_p$ with a $G$-homotopy equivalent chain complex. Furthermore, there is a bijection between projective modules over $\mathbb{F}_p$ and over $\mathbb{Z}_p$, allowing us to work with $\mathbb{F}_p$-coefficients. Each term in the Steinberg complex is a sum of permutation modules, where the stabilisers are the chain normalisers, i.e. if $\sigma = (Q_0 < \cdots < Q_s) \in C_s(|\mathcal{S}_p(G)^*|; \mathbb{Z}_p)$, then $G_\sigma = \bigcap\limits_{0 \le i \le s} N_G(Q_i)$. In Theorem 5.5, the Steinberg complex is *twisted*, in the sense that instead of taking trivial coefficients, we have a one dimensional module $k_\varphi$. (We defined $G$-twisted coefficient systems in Sect. 1.6.3.)

Now, let $G_0 = \langle N_G(Q) \mid 1 < Q \le S \rangle$ be the smallest strongly $p$-embedded subgroup of $G$. So $K(G) \cong K(G_0)$ by Proposition 3.3. [70, Theorem B] relates $K(G)$ to the $p$-subgroup structure of $G_0$.

**Theorem 5.6.** *There exists an exact sequence*

$$0 \longrightarrow \mathrm{Hom}(G_0, k^{\times}) \longrightarrow K(G) \longrightarrow \mathrm{H}^1(\mathscr{S}_p^*(G_0); k^{\times})^{G_0} \longrightarrow \mathrm{H}^2(G_0; k^{\times})$$

*where the superscript $G_0$ indicates the $G_0$-invariants.*

*In particular, if the $G$-coinvariants $\mathrm{H}_1(\mathscr{S}_p^*(G); k^{\times})_G$ are zero, then $K(G) \cong G_0/G_0 S'$, and if $\pi_1(|\mathscr{S}_p^*(G)|) = 0$, then $K(G) \cong G/G'S$.*

Theorem 5.5 leads to the interpretation of $K(G)$ using the Borel construction $|\mathscr{S}_p^*(G)|_{\mathrm{h}\,G}$ of the poset of nontrivial $p$-subgroups of $G$, which is homotopy equivalent to the nerve $|\mathscr{T}_p^*(G)|$ of the transport category. Since $\mathscr{T}_p^*(G)$ is homotopy equivalent to the orbit category $\mathscr{O}_p^*(G)$, the group $K(G)$ is isomorphic to $\mathrm{Hom}(\pi_1(\mathscr{T}_p^*(G)), k^{\times})$, and [70, Corollary C] follows.

**Corollary 5.1.** *There is an isomorphism $K(G) \cong \mathrm{H}^1(|\mathscr{S}_p^*(G)|_{\mathrm{h}\,G}; k^{\times})$. In particular, if $\mathrm{H}_1(|\mathscr{S}_p^*(G)|_{\mathrm{h}\,G}; k^{\times}) \cong \mathrm{H}_1(G; k^{\times})$, then $K(G) \cong G/G'S$.*

When it comes to computations, the descriptions of $K(G)$ using homology decompositions are the most useful. These are the normaliser and the centraliser decompositions.

**Theorem 5.7.** ([70, Theorem D] *Let $\mathscr{S} \subseteq \mathscr{S}_p^*(G)$ be an ample collection of nontrivial $p$-subgroups of $G$ (cf. Sect. 1.6). Then*

$$K(G) \cong \varprojlim_{\sigma \in C^*(\mathscr{S})} \mathrm{H}^1(N_G(\sigma); k^{\times}),$$

*where the limit is taken over all the chains in $\mathscr{S}$ ordered by refinement. Explicitly,*

$$K(G) \cong \ker \Big( \bigoplus_{[Q] \in C^0(\mathscr{S})} \mathrm{H}^1(N_G(Q); k^{\times}) \to \bigoplus_{[Q<R] \in C^1(\mathscr{S})} \mathrm{H}^1(N_G(Q) \cap N_G(R); k^{\times}) \Big).$$

**Theorem 5.8.** ([70, Theorem E]) *There exists an exact sequence*

$$0 \to \mathrm{H}^1(\mathscr{F}_p^*(G); k^{\times}) \to K(G) \to \varprojlim_{E \in \mathscr{F}_{\mathscr{E}^*}(G)} \mathrm{H}^1(C_G(E); k^{\times}) \to \mathrm{H}^2(\mathscr{F}_p^*(G); k^{\times}),$$

*where $\mathscr{E}^*$ denotes the poset of nontrivial elementary abelian $p$-subgroups of $G$. In particular, if $\mathrm{H}_1(C_G(x); k^{\times}) = 0$ for each $x \in G$ of order $p$, and if $N_G(S)$ is generated by $N_G(S)'S$ and by elements which centralise some nontrivial element of $S$, then $K(G) = 0$.*

Theorem 5.7 implies that Carlson and Thévenaz's conjecture [45, Conjecture 5.5] holds. In fact, [70, Theorem F] gives a sharper result.

**Theorem 5.9.** *Consider the $\rho^j(S)$ subgroups from Definition 5.2. There is a group isomorphism*

$$H_1(\mathscr{O}_p^*(G); k^\times) \cong N_G(S)/\rho^r(S),$$

*for any $r > \dim |\mathscr{B}_p^*(G)|$. It follows by Theorem 5.5 that for any $r > \dim |\mathscr{B}_p^*(G)|$, we have $K(G) \cong \mathrm{Hom}(N_G(S)/\rho^r(S), k^\times)$.*

Here $\mathscr{B}_p^*(G)$ is the poset of nontrivial $p$-radical subgroups of $G$ (cf. Sect. 1.2), and $\dim |\mathscr{B}_p^*(G)|$ is the maximum of the lengths of the chains of nontrivial $p$-radical subgroups of $G$, ordered by inclusion.

As a consequence of Theorem 5.9, if $S$ is subject to some conditions, then we obtain [70, Corollary G].

**Corollary 5.2.** *Suppose that every nontrivial $p$-radical subgroup of $G$ contained in $S$ is normal in $S$. Then*

$$K(G) \cong \ker\left( H^1(N_G(S); k^\times) \to \bigoplus_{[Q]} H^1\left((N_G(Q)/N_G(Q)'S) \cap N_G(S); k^\times\right)\right),$$

*where $[Q] \in \mathscr{B}_p^*(G)$ runs through the $N_G(S)$-conjugacy classes of proper nontrivial $p$-radical subgroups of $S$. In particular, $K(G) \cong \mathrm{Hom}(N_G(S)/\rho^2(S), k^\times)$.*

The last general result, [70, Theorem H], draws on $p$-fusion properties. Write $\mathscr{O}_p^c(G)$ for the orbit category on the collection of $p$-centric subgroups of $G$, and recall that $\mathrm{Out}_G(S) = N_G(S)/SC_G(S)$.

**Theorem 5.10.** *There is a commutative diagram of monomorphisms*

$$
\begin{array}{ccccc}
K(G) & \longrightarrow & \mathrm{Hom}(\pi_1(\mathscr{O}_p^c(G), k^\times) & \longrightarrow & \mathrm{Hom}(N_G(S)/S, k^\times) \\
\uparrow & & \uparrow & & \uparrow \\
\mathrm{Hom}(\pi_1(\mathscr{F}_p^*(G)), k^\times) & \longrightarrow & \mathrm{Hom}(\pi_1(\mathscr{F}_p^c(G)), k^\times) & \longrightarrow & \mathrm{Hom}(\mathrm{Out}_G(S), k^\times).
\end{array}
$$

*In particular, if all the $p$-centric $p$-radical subgroups are centric, then*

$$\mathrm{Hom}\left(\pi_1(\mathscr{F}_p^*(G)), k^\times\right) \le K(G) \le \mathrm{Hom}\left(\pi_1(\mathscr{F}_p^c(G)), k^\times\right).$$

The final theorem in [70, Theorem I] is an application of Theorem 5.6 to the finite symplectic groups subject to certain constraints.

**Theorem 5.11.** *Let $G = \mathrm{Sp}_{2n}(q)$ for a prime power $q$. Suppose that the multiplicative order of $q$ modulo $p$ is odd, and that $G$ has $p$-rank at least 3, then $K(G) = 0$.*

We will see more "successful" applications of Grodal's homotopy method in Sect. 6.2.

# Chapter 6
# Endotrivial Modules for Very Important Groups

In Sects. 3.3, 3.4 and 3.6 through Sect. 3.11, we give the classification of endotrivial modules for many finite groups: of prime power order, with a normal Sylow
$p$-subgroup, with a cyclic Sylow $p$-subgroup for any prime, with a dihedral, quaternion or semi-dihedral Sylow 2-subgroup, and finally for finite $p$-soluble groups. By
*classification*, we mean the isomorphism type of the group $T(G)$, and, if known, a
presentation of $T(G)$ by generators and relations.

In this chapter we state the results known to date about the structure of the group
of endotrivial modules for large classes of finite groups: symmetric and alternating
groups and their covering groups, finite groups of Lie type in defining and nondefining characteristic and sporadic groups and their covering groups.

## 6.1 Symmetric and Alternating Groups and Their Covering Groups

The classification of endotrivial modules for symmetric and alternating groups was
obtained in [35] in the case of characteristic $p = 2$ and for odd primes in the case
when a Sylow $p$-subgroup is abelian. The result was then extended to all symmetric
and alternating groups for odd primes in [33]. Finally, in [91], the authors determined
$T(G)$ for the covering groups of all the symmetric and alternating groups.

Leaving aside a few exceptions, the influencing factors which determine the structure of $T(G)$ are mainly of two strands:

- $p = 2$ versus $p$ odd, and
- abelian versus nonabelian Sylow $p$-subgroups.

Let $G$ be a symmetric or alternating group, and let $S \in \mathrm{Syl}_p(G)$. If $S$ is abelian,
then $S$ is elementary abelian of rank at most $p - 1$, in which case we distinguish
between cyclic, rank 2 and larger ranks. If $S$ is nonabelian, then $S$ is a direct product

© The Author(s), under exclusive license to Springer Nature Switzerland AG 2019          87
N. Mazza, *Endotrivial Modules*, SpringerBriefs in Mathematics,
https://doi.org/10.1007/978-3-030-18156-7_6

of iterated wreath products of groups of order $p$. If $p = 2$, the situation is easier, for the Sylow 2-subgroups are selfnormalising, except for the small groups. In any case, the $p$-local group structure of symmetric and alternating groups is well-known, and has been used in determining the structure of $T(G)$. The only question open to date is to find a suitable set of generators for $TF(G)$ when $TF(G)$ is noncyclic.

### 6.1.1   Results and Method for $\mathfrak{S}_n$ and $A_n$

Let us summarise the main theorems of [33, 35] for the symmetric and for the alternating groups.

**Theorem 6.1.** *Let $\mathfrak{S}_n$ be the symmetric group on $n$ letters.*

1. *If $p = 2$, then*

$$T(\mathfrak{S}_n) \cong \begin{cases} \{0\} & if\, n \leq 3, \\ \mathbb{Z}^2 & if\, n = 4,\, 5, \\ \mathbb{Z} & if\, n \geq 6. \end{cases}$$

2. *If $p \geq 3$, then*

$$T(\mathfrak{S}_n) \cong \begin{cases} \{0\} & if\, n < p, \\ \mathbb{Z}/2(p-1) & if\, n = p,\ p+1, \\ \mathbb{Z}/2(p-1) \oplus \mathbb{Z}/2 & if\, p+2 \leq n < 2p, \\ \mathbb{Z} \oplus (\mathbb{Z}/2)^2 & if\, 2p \leq n < 3p, \\ \mathbb{Z} \oplus \mathbb{Z}/2 & if\, 3p \leq n < p^2, \\ \mathbb{Z}^2 \oplus \mathbb{Z}/2 & if\, p^2 \leq n < p^2 + p, \\ \mathbb{Z} \oplus \mathbb{Z}/2 & if\, p^2 + p \leq n, \end{cases}$$

*where $K(\mathfrak{S}_n) \cong \mathrm{Hom}(\mathfrak{S}_n, k^\times)$ is generated by the stable isomorphism classes of one-dimensional modules, except when $n < 2p$, in which case $\mathfrak{S}_n$ has a cyclic Sylow $p$-subgroup, and when $2p \leq n < 3p$, in which case the stable isomorphism class $[Y^{(n-p,p)}]$ of the Young module labeled by the partition $(n - p, p)$ of $n$ is selfdual endotrivial.*

**Theorem 6.2.** *Let $A_n$ be the alternating group on $n$ letters.*

1. *If $p = 2$, then*

$$T(A_n) \cong \begin{cases} \{0\} & if\, n \leq 3, \\ \mathbb{Z} \oplus \mathbb{Z}/3 & if\, n = 4,\, 5, \\ \mathbb{Z}^2 & if\, n = 6,\, 7, \\ \mathbb{Z} & if\, n \geq 8. \end{cases}$$

2. *If $p \geq 3$, then*

$$T(A_n) \cong \begin{cases} \{0\} & if\ n < p, \\ \mathbb{Z}/(p-1) \oplus \mathbb{Z}/2 & if\ n = p,\ p+1, \\ \mathbb{Z}/2(p-1) & if\ p+2 \le n < 2p, \\ \mathbb{Z} \oplus \mathbb{Z}/4 & if\ p \ge 3\ \text{and}\ n = 2p,\ 2p+1, \\ \mathbb{Z} \oplus \mathbb{Z}/2 & if\ 2p+2 \le n < 3p, \\ \mathbb{Z} & if\ 3p \le n < p^2,\ \text{and if}\ p^2 + p \le n, \\ \mathbb{Z}^2 & if\ p^2 \le n < p^2 + p. \end{cases}$$

To prove these theorems, we start by analysing the $p$-local structure of $\mathfrak{S}_n$, and then intersecting with $A_n$ to study that of the alternating groups.

For $n \in \mathbb{N}$, write $n = \sum_{j \ge 0} a_j p^j$ with $0 \le a_j < p$. Then

$$S \cong \prod_{j \ge 0} (C_p^{ij})^{a_j} \in \mathrm{Syl}_p(\mathfrak{S}_n), \text{ where } C_p^{ij} = (C_p^{i(j-1)}) \wr C_p$$

is a direct product of iterated wreath products defined inductively, beginning with $C_p^{i0} = 1$ and $C_p^{i1} = C_p$. In particular, $S$ is abelian if and only if $S$ is elementary abelian of rank less than $p$, and $S$ is a TI subgroup of $G$ whenever $S$ is cyclic. For $A_n$, we note that $S = S_A \in \mathrm{Syl}_p(A_n)$, unless $p = 2$, in which case $S_A = S \cap A_n \in \mathrm{Syl}_2(A_n)$ has index 2 in $S$.

From [33, Sect. 4], with $G = \mathfrak{S}_n$,

$$N_G(S) \cong \prod_{j \ge 0} (N_j \wr \mathfrak{S}_{a_j}), \text{ where } N_j = N_{\mathfrak{S}_{p^j}}(C_p^{ij}) \cong C_p^{ij} \rtimes C_{p-1}^j.$$

Consequently, if $p = 2$ and $S$ is noncyclic, then $S$ is selfnormalising.

For $A_n$, note that $N_{A_n}(S_A)$ has index 2 in $N_G(S)$ for all primes except if $n = 4, 5$ and $p = 2$, in which case $N_{A_n}(S_A) \cong A_4$.

Moreover, $\mathfrak{S}_4$, $\mathfrak{S}_5$, $A_6$ and $A_7$ have selfnormalising dihedral Sylow 2-subgroups of order 8, giving $T(G) \cong \mathbb{Z}^2$, as described in Sect. 3.8.

When $S = N_G(S)$, we know $T(S)$ from the classification of endotrivial modules for finite $p$-groups (Sect. 3.3), and the rest of the assertion for $p = 2$ follows since $T(S) = \langle \Omega_S \rangle$ for most $n$.

So, we are left with $p$ odd. Then, we proceed case by case, using known results and direct computations with the algebra software Magma [15].

We start with a useful observation on the restriction map if $p$ is odd and $S$ noncyclic.

**Lemma 6.1.** ([35, Lemma 3.2]) *Suppose that $n > 2p$ and that $p \nmid n$. Then he restriction maps* $\mathrm{Res}_{\mathfrak{S}_{n-1}}^{\mathfrak{S}_n} : T(\mathfrak{S}_n) \to T(\mathfrak{S}_{n-1})$ *and* $\mathrm{Res}_{A_{n-1}}^{A_n} T(A_n) \to T(A_{n-1})$ *are injective. Moreover, if $p$ is odd, then* $\ker(\mathrm{Res}_{A_n}^{\mathfrak{S}_n})$ *is generated by the stable isomorphism class of the sign representation, which has order 2 in* $T(\mathfrak{S}_n)$.

*Proof (of Theorem 6.1).* We only outline the argument for $G = \mathfrak{S}_n$, $p$ odd and $S$ noncyclic. Suppose that $n \ge 2p$. Then $T(G) = K(G) \oplus TF(G)$ since $T(S)$ is torsionfree. The rank of $TF(G)$ is equal to the number of connected components of the poset $\mathscr{E}_{\geq 2}(G)/G$ of non-cyclic elementary abelian $p$-subgroups of $G$, which is equal to 1 if $2p \leq n < p^2$ or $p^2 + p \leq n$, and is equal to 2 if $p^2 \leq n < p^2 + p$.

Thus, the key problem is to find $K(G)$. By definition, any trivial Sylow restriction module must be some Young module. To determine which Young modules are endotrivial, we proceed by "$p$-power step induction", that is, we start with the case $n = ap$ for some $a \geq 2$. We need to take the 1-dimensional $kN_G(S)$-modules, knowing that $K(N_G(S)) \cong N_G(S)/N_G(S)'S \cong C_{p-1} \times C_2$. Then we show that the restriction of an endotrivial Young module to the subgroup $H = N_G(S)\mathfrak{S}_{(p,\dots,p)}$ of $G = \mathfrak{S}_{ap}$ must be the direct sum of a 1-dimensional $kH$-module and a projective module. Here, $\mathfrak{S}_\lambda$ for a partition $\lambda = (\lambda_1, \dots, \lambda_t)$ of $m \in \mathbb{N}$ is the Young subgroup $\mathfrak{S}_{\lambda_1} \times \cdots \times \mathfrak{S}_{\lambda_t}$ of $\mathfrak{S}_m$.

If $n = 2p$, then $H$ is strongly $p$-embedded in $G$ and the result follows.

If $2p < n < 3p$, then we use A. Henke's computations of $p$-Kotska numbers in [77] and Lemma 6.1 to prove that the only endotrivial Young modules are those labeled by the partitions $(n - p, p)$.

If $3p \leq n < p^2$, then we use Mackey's formula with well chosen Young subgroups of $G$ to see that the only nontrivial indecomposable endotrivial Young module is the sign representation.

To prove the theorem in the case when $S$ is not abelian, we continue with a "$p$-power step induction". The critical step is $n = p^2$, for which we show that $K(G) \cong G/G'S$ using our result for $\mathfrak{S}_{p^2-1}$, Mackey's Formula and *Frobenius reciprocity*, which says that given a subgroup $H$ of a finite group $G$, a $kG$-module $M$ and a $kH$-module $N$, then $\operatorname{Hom}_{kG}(M, N\uparrow_H^G) \cong \operatorname{Hom}_{kH}(M\downarrow_H^G, N)$ (cf. [11, Volume I, Proposition 3.3.1]). For the alternating groups, we show that the restriction $\operatorname{Res}_{A_n}^{\mathfrak{S}_n}(Y^{(\lambda|\mu)})$ of certain *signed Young modules* $Y^{(\lambda|\mu)}$ for $\mathfrak{S}_n$ produces nontrivial indecomposable trivial Sylow restriction $kA_n$-modules for $n = 2p, 2p + 1$.    □

### 6.1.2 Results and Method for the Covering Groups of $\mathfrak{S}_n$ and $A_n$

**Definition 6.1.** Let $G$ be a finite group. A *covering group* of $G$, or *Schur cover*, is a finite group $\tilde{G}$ such that $Z(\tilde{G}) \leq \tilde{G}'$ and $\tilde{G}/Z(\tilde{G}) \cong G$. It is known that a finite perfect group has a unique maximal covering group, which we call the *universal cover*.

Two groups $G$ and $H$ are *isoclinic* if there exist group isomorphisms $\varphi : G/Z(G) \to H/Z(H)$ and $\tau : G' \to H'$ such that the diagram

$$
\begin{array}{ccc}
G/Z(G) \times G/Z(G) & \xrightarrow{\gamma_G} & G' \text{ commutes,} \\
\downarrow{\varphi} & & \downarrow{\tau} \\
H/Z(H) \times H/Z(H) & \xrightarrow{\gamma_H} & H'
\end{array}
$$

where $\gamma_K (x Z(K), y Z(K)) = [x, y]$ for any $x, y \in K$ and any group $K$.

Detailed constructions of the covering groups of the alternating and symmetric groups can be found in [80, Chap. 2] and in [126, Sect. 2.7.2]. For $n \geq 4$ the Schur multiplier of the symmetric and alternating groups is nontrivial of order 2, except in the cases of $A_6$ and $A_7$, in which case it has order 6. There are two non-isomorphic isoclinic covers for the symmetric groups, except if $n = 6$ when both are isomorphic. There is a unique maximal cover for the alternating groups. Following the standard convention, we let $2^{\pm}.\mathfrak{S}_n$ denote the two different isoclinic double covers of the symmetric group $\mathfrak{S}_n$, and $d.A_n$ the $d$-fold cover of the alternating group $A_n$. So, $2^+.\mathfrak{S}_n$ is the cover of $\mathfrak{S}_n$ in which transpositions of $\mathfrak{S}_n$ lift to involutions, and in $2^-.\mathfrak{S}_n$ transpositions lift to elements of order 4.

The presentation by generators and relations of $2^{\pm}.\mathfrak{S}_n$ is

$$2^{\pm}.\mathfrak{S}_n = \langle z, t_1, \ldots, t_{n-1} \rangle / \langle \mathscr{R} \rangle \text{ (cf. [80])},$$

where $\mathscr{R}$ is generated by the relations

$$z^2 = 1,$$
$$t_j^2 = z^{\alpha}, \quad 1 \leq j \leq n - 1,$$
$$(t_j t_{j+1})^3 = z^{\alpha}, \quad 1 \leq j \leq n - 2,$$
$$[z, t_j] = 1, \quad 1 \leq j \leq n - 1,$$
$$[t_j, t_k] = z, \text{ if } |j - k| > 1 \text{ for all } 1 \leq j, k \leq n - 1,$$

where $\alpha = 0$ for $G = 2^+.\mathfrak{S}_n$ and $\alpha = 1$ for $G = 2^-.\mathfrak{S}_n$. Then $2.A_n$ is the preimage of $A_n$ in any of $2^{\pm}.\mathfrak{S}_n$ via the natural quotient map $G \mapsto G/Z(G)$, where $Z(G) = \langle z \rangle$. It follows that

$$2.A_n = \langle z, x_1, \ldots, x_{n-2} \rangle / \left( \langle \mathscr{R} \rangle \cap \langle z, x_1, \ldots, x_{n-2} \rangle \right)$$

where $x_j = t_j t_{j+1}$ for $1 \leq j \leq n - 2$.

The $p$-local structures of $2.\mathfrak{S}_n^{\pm}$ and $d.A_n$ are well-known, and, in particular, the Sylow 2-subgroups are selfnormalising. It follows that $T(G)$ for the covering groups of alternating and symmetric groups in characteristic 2 is immediate. (Note that the groups $2.\mathfrak{S}_n^{\pm}$ and $2.A_n$ have a nontrivial normal 2-subgroup.) So finding $T(G)$ amounts to calculating the number of connected components of $\mathscr{E}_{\geq 2}(G)/G$.

**Theorem 6.3.** ([91, Theorem A]) *Suppose that $p = 2$, and that k contains a primitive cube root of 1. The following hold.*

1. $T(2.A_n) \cong \begin{cases} \mathbb{Z}/2 \oplus \mathbb{Z}/4 & \text{if } 4 \leq n \leq 7, \\ \mathbb{Z} & \text{if } 8 \leq n. \end{cases}$

2. $T(2^+.\mathfrak{S}_n) \cong \begin{cases} \mathbb{Z}/2 \oplus \mathbb{Z} & \text{if } n \in \{4, 5\}, \\ \mathbb{Z}^2 & \text{if } n \in \{8, 9\}, \\ \mathbb{Z} & \text{if } n \in \{6, 7\} \text{ or } 10 \leq n. \end{cases}$

3. $T(2^-.\mathfrak{S}_n) \cong \begin{cases} \mathbb{Z}/2 \oplus \mathbb{Z}/4 \; \text{if} \; n \in \{4, 5\}, \\ \mathbb{Z} \qquad\qquad \text{if} \; 6 \leq n. \end{cases}$

If $p$ is odd, then building on [33, 35] and using the methods from [89], we obtain a complete description of $T(G)$ for $p$ odd.

**Theorem 6.4.** ([91, Theorem B]) *Suppose that $p \geq 3$. Let $G$ be one of the double covers $2.A_n$ or $2^\pm.\mathfrak{S}_n$ with $n \geq 4$. The following hold.*

1. *If $p = 3$ and $n = 6$, then $T(2.A_6) \cong \mathbb{Z}^2 \oplus \mathbb{Z}/8$, and $T(2.\mathfrak{S}_6) \cong \mathbb{Z} \oplus (\mathbb{Z}/2)^2$.*
2. *If $n \geq p + 4$, then the inflation $\mathrm{Inf}^G_{G/Z(G)} : T(G/Z(G)) \to T(G)$ is an isomorphism.*
3. *If $p \geq 3$ and $p \leq n \leq \min\{2p - 1, p + 3\}$ then $T(G) \cong \mathbb{Z}/2e \oplus \mathbb{Z}/2$, where $e$ denotes the inertial index of the principal block of $kG$.*

Note that in Theorem 6.4, we have $T(2.\mathfrak{S}_6) = \mathrm{Inf}^{2.\mathfrak{S}_6}_{\mathfrak{S}_6}\left(T(\mathfrak{S}_6)\right)$. As expected, for $p$ odd, the group of endotrivial modules for the covering groups of the alternating and symmetric groups is often equal to the image of the inflation map.

We end with the classification theorem for the remaining covering groups of $A_6$ and $A_7$ as obtained in [91, Sect. 6].

**Theorem 6.5.** *Let $k$ be a field of characteristic $p$, and suppose that $k$ is large enough for the groups considered.*

1. *If $p = 2$, then*

$$T(3.A_6) \cong \mathbb{Z}^2 \oplus \mathbb{Z}/3,$$
$$T(6.A_6) \cong \mathbb{Z}/4 \oplus \mathbb{Z}/2,$$
$$T(3.A_7) \cong \mathbb{Z}^2,$$
$$T(6.A_7) \cong \mathbb{Z}/4 \oplus \mathbb{Z}/2.$$

2. *If $p = 3$, then $T(G) = \langle \Omega_G, [M] \rangle \cong \mathbb{Z}^2$, where $M$ is a $kG$-module such that*

$$\mathrm{Res}^G_F(M) \cong \Omega^6_F(k) \oplus (\mathrm{proj}) \quad \text{and} \quad \mathrm{Res}^G_E(M) \cong k \oplus (\mathrm{proj})$$

*for some maximal elementary abelian 3-subgroup $F$ of $G$ of rank 2 and for any elementary abelian 3-subgroup $E$ of rank 2 of $G$ not conjugate to $F$.*
3. *If $p = 5$, then*

$$T(3.A_6) \cong \mathbb{Z}/4 \oplus \mathbb{Z}/3,$$
$$T(6.A_6) \cong \mathbb{Z}/4 \oplus \mathbb{Z}/3 \oplus \mathbb{Z}/2,$$
$$T(3.A_7) \cong \mathbb{Z}/8 \oplus \mathbb{Z}/3,$$
$$T(6.A_7) \cong \mathbb{Z}/8 \oplus \mathbb{Z}/3 \oplus \mathbb{Z}/2.$$

4. *If $p = 7$, then*

$$T(3.A_7) \cong (\mathbb{Z}/3)^2 \oplus \mathbb{Z}/2,$$
$$T(6.A_7) \cong \mathbb{Z}/6 \oplus \mathbb{Z}/3 \oplus \mathbb{Z}/2.$$

## 6.2 Finite Groups of Lie Type

Finite groups of Lie type are ubiquitous in the classification of finite simple groups. The literature about these groups is abundant, but there is a proliferation of definitions with subtle variations which can be very confusing. So our first task in this section is to say what we mean by a *finite group of Lie type*. We start with some concepts from Lie theory and algebraic groups, for which we refer to [98] (in particular [98, Sects. 8, 9, 21 and 22]).

Since we need a minuscule fraction of the vast theory on groups of Lie type, we only review the relevant notions that we will use in the sequel, assuming some elementary knowledge about linear algebraic groups. In particular, we refer to [98, Definition 6.15] for the definitions of (semi-)simple and reductive linear algebraic groups, and to [98, Sects. 8 and 9] for a detailed discussion of reductive groups and root data.

Let $\mathbf{G}$ be a linear algebraic group defined over an algebraically closed field of characteristic $\ell$. A *Steinberg endomorphism* of $\mathbf{G}$ is an endomorphism of algebraic groups $F : \mathbf{G} \to \mathbf{G}$ such that there exists some positive integer $m$ for which $F^m : \mathbf{G} \to \mathbf{G}$ is the *Frobenius morphism* with respect to some $\ell$-power $q$. We call $F$ $\mathbb{F}_q$-*split* if there exists a maximal $F$-stable torus of $\mathbf{G}$ on which $F$ acts by raising the elements to the $q$-th power. $F$ is *twisted* if $F$ is not split but the product of an $\mathbb{F}_q$-split endomorphism with an automorphism of algebraic groups. $F$ is *very twisted* if some power of $F$ induces a nontrivial graph automorphism (i.e. if $\mathbf{G}$ is irreducible, then $(\mathbf{G}, F)$ has type $^2B_2$, $^2G_2$ or $^2F_4$).

*Remark 6.1.* Throughout this section, the notation $F$ refers to a Steinberg endomorphism, not the field of fractions in a $p$-modular system (with residue field $k$).

**Definition 6.2.** Let $\mathbf{G}$ be a semisimple algebraic group defined over an algebraically closed field of characteristic $\ell$. Let $F$ be a Steinberg endomorphism of $\mathbf{G}$, and write $G = \mathbf{G}^F$ for the finite group obtained by taking $F$-fixed points.

1. We call $G$ a *finite group of Lie type*.
2. Now, fix an $F$-stable maximal torus $\mathbf{T}$ of $\mathbf{G}$. Write $\mathbf{R}$ for the set of *roots* and $W = N_{\mathbf{G}}(\mathbf{T})/C_{\mathbf{G}}(\mathbf{T})$ for the *Weyl group* of $\mathbf{G}$ with respect to $\mathbf{T}$. That is, $\mathbf{R}$ is the set of nonzero characters $X$ of $\mathbf{T}$ with nonzero eigenspace. Write $Y$ for the group of cocharacters of $\mathbf{T}$ and $\mathbf{R}^\vee$ for the coroots. Each 5-tuple $\mathbf{G}, \mathbf{T}, W, \mathbf{R}, X$ defines a *root datum*, namely the quadruple $\mathbf{D} = (X, \mathbf{R}, Y, \mathbf{R}^\vee)$. It is subject to the following axioms.

a.  $X \cong \mathbb{Z}^n \cong Y$ and there is a bilinear map $\langle *, * \rangle : X \times Y \to \mathbb{Z}$ such that any homomorphism $X \to \mathbb{Z}$ or $Y \to \mathbb{Z}$ is of the form $x \mapsto \langle x, y_0 \rangle$ or $y \mapsto \langle x_0, y \rangle$ for some $x_0 \in X$ and $y_0 \in Y$.

b.  $\mathbf{R} \subseteq X$ and $\mathbf{R}^\vee \subseteq Y$ are abstract root systems in the $\mathbb{R}$-vector spaces $\mathbb{Z}\mathbf{R} \otimes_{\mathbb{Z}} \mathbb{R}$ and $\mathbb{Z}\mathbf{R}^\vee \otimes_{\mathbb{Z}} \mathbb{R}$, respectively.

c.  There exists a bijection $\mathbf{R} \to \mathbf{R}^\vee$ such that $\langle \alpha, \alpha^\vee \rangle = 2$ for each $\alpha \in \mathbf{R}$.

d.  Each (co-)root $\alpha$ (resp. $\alpha^\vee$) gives a reflection $s_\alpha$ (resp. $s_{\alpha^\vee}$) of the (co-)root system such that $s_\alpha(x) = x - \langle x, \alpha^\vee \rangle$ and $s_{\alpha^\vee}(y) = y - \langle \alpha, y \rangle$ for all $x \in X, y \in Y$ and $\alpha \in \mathbf{R}$. Thus $W$ is generated by all such $s_\alpha$ (or equivalently up to isomorphism, all the $s_{\alpha^\vee}$).

e.  Let $Y' = \mathrm{Hom}(\mathbb{Z}\mathbf{R}^\vee, \mathbb{Z})$. Restriction along the inclusion $\mathbb{Z}\mathbf{R}^\vee \hookrightarrow Y$ induces an injective group homomorphism

$$\underbrace{\mathrm{Hom}(Y, \mathbb{Z})}_{\cong X} \longrightarrow Y'$$

and hence inclusions $\mathbb{Z}\mathbf{R} \subseteq X \subseteq Y'$. The quotient group $Y'/\mathbb{Z}\mathbf{R}$ is called the *fundamental group of the root system* $\mathbf{R}$. If $X = Y'$, then $\mathbf{G}$ is *simply connected*, and if $X = \mathbb{Z}\mathbf{R}$, then $\mathbf{G}$ is of *adjoint* type (cf. [98, Table 22.1]).

3.  A *complete root datum* is a root datum $(X, \mathbf{R}, Y, \mathbf{R}^\vee)$ together with a coset $W\tau$ of $W$ in the semidirect product $W \rtimes \langle F \rangle$. It follows that the pair $(\mathbf{G}, F)$ is determined up to isomorphism by the complete root datum $(X, \mathbf{R}, Y, \mathbf{R}^\vee, W\tau)$ and an $\ell$-power $q$ such that $F$ acts as $q\delta$ on $X_{\mathbb{R}} = \mathbb{R} \otimes_{\mathbb{Z}} X$, where $\delta$ is the order of the graph automorphism of the corresponding Dynkin diagram ([98, Theorem 22.5]). We then say that $\mathbf{G}$ *is defined over* $\mathbb{F}_q$. Each root system has a *basis of simple roots*, whose cardinality is the *Lie rank* of $\mathbf{G}$. If $\delta > 1$, then $\tau$ acts on $X_{\mathbb{R}}$, and we let $\tilde{R}$ be the image of $R$ under the projection of $X_{\mathbb{R}}$ onto $C_{X_{\mathbb{R}}}(\tau)$. The set of equivalence classes $\hat{R} = \tilde{R}/\sim$ with respect to the relation: $\tilde{x} \sim \tilde{x}'$ *if and only if* $\tilde{x} = c\tilde{x}'$ *for some positive number* $c$ is the *twisted root system*, whose cardinality is the *twisted Lie rank* of $(\mathbf{G}, F)$ (cf. [68, Definition 2.3.1]).

4.  The order of $G$ is the evaluation at $q$ of a certain polynomial, called the *polynomial order* $|(\mathbf{G}, F)| \in \mathbb{Z}[x]$ of $(\mathbf{G}, F)$, which only depends on $\mathbf{D}$. Now $|(\mathbf{G}, F)|$ can be factorised as a product of a power of $x$ and cyclotomic polynomials $\Phi_j \in \mathbb{Z}[x]$. The $\Phi_d$-*rank* of $(\mathbf{G}, F)$ and of $G$ is the multiplicity of $\Phi_d$ in $|(\mathbf{G}, F)|$.

The celebrated classification theorem of Chevalley asserts that two semisimple linear algebraic groups are isomorphic if and only if they have isomorphic root data, and for each root datum there is a semisimple algebraic group $\mathbf{G}$ that realises it. Moreover, $\mathbf{G}$ is simple if its root datum is indecomposable ([98, Theorem 9.13]).

The description of the $p$-local structure of $G$ when $\ell \neq p$ relies on the following Lie-theoretic gadget.

**Definition 6.3.** Let $\mathbf{G}$ and $\mathbf{T}$ be as in Definition 6.2.

1. An $F$-stable torus $\mathbf{T}_d$ is a $\Phi_d$-*torus* if the polynomial order $|\mathbf{T}_d|$ of $\mathbf{T}_d$ is a power of the $d$-th cyclotomic polynomial $\Phi_d$.
2. We say that $\mathbf{T}_d$ is *maximal*, if $|\mathbf{T}_d| = |\mathbf{G}|_{\Phi_d}$, i.e. the polynomial order of $\mathbf{T}_d$ is equal to the $\Phi_d$-part of $|\mathbf{G}|$.
3. The centraliser of a $\Phi_d$-torus is called a $d$-*split Levi subgroup*.
4. The group $W_d$ is the *complex reflection group* $W_d = N_G(\mathbf{T}_d)/\mathbf{T}_d^F$, where $\mathbf{T}_d$ is a maximal $\Phi_d$-torus of $\mathbf{G}$.

In particular, $\Phi_d$-tori are $F$-stable tori of $\mathbf{G}$, and they have numerous properties which we will use later. We record here the main ones (cf. [98, Theorem 25.11 and Corollary 25.17]). Most of these can be extended to connected reductive groups.

**Theorem 6.6.** *Let $\mathbf{G}$ be a connected semisimple algebraic group defined over an algebraically closed field of characteristic $\ell$. Let $F$ be a Steinberg endomorphism of $\mathbf{G}$, and write $G = \mathbf{G}^F$ for the finite group obtained by taking $F$-fixed points, where $F^m : \mathbf{G} \to \mathbf{G}$ is the Frobenius morphism with respect to $q = \ell^a$. Let $d$ be a positive integer, and suppose the same notation as in Definition 6.3.*

1. *Maximal $\Phi_d$-tori exist in $\mathbf{G}$ and they are all $\mathbf{G}$-conjugate.*
2. *Any $\Phi_d$-torus is contained in a maximal $\Phi_d$-torus.*
3. *Suppose that $p$ is a prime different from $\ell$ and $d$ is the multiplicative order of $q$. If $p = 2$, put $d = 1$ if $\ell \equiv 1 \pmod 4$ and $d = 2$ if $\ell \equiv 3 \pmod 4$. If $F$ is not very twisted and $p \geq 5$, then each Sylow $p$-subgroup of $G$ is contained in some $N_G(\mathbf{T}_d)$, for some maximal $\Phi_d$-torus of $\mathbf{G}$.*

Now, $\Phi_d$-tori are $F$-stable tori, and therefore, [98, Corollary 8.13] says that their centralisers are connected reductive algebraic groups. But by contrast with maximal tori, the maximal $\Phi_d$-tori may not be selfcentralising, and we shall come back to this fact in the study of $K(G)$.

Let us now exhibit the different $p$-local structures that we can find in an example.

*Example 6.1.* The well known paradigm of a finite group of Lie type is $G = \mathrm{SL}_n(q)$, for $q = \ell^a$ and $n \geq 2$ (to avoid trivialities). So $\mathbf{G} = \mathrm{SL}_n$ and the Steinberg endomorphism is given by the Frobenius map $x \mapsto x^q$ for $x \in \overline{\mathbb{F}}_\ell$. This group is simply connected of type $A_{n-1}$, of (twisted) Lie rank $n - 1$. (By comparison, the twisted Lie rank of $\mathrm{SU}_n(q)$ is $\lfloor \frac{n}{2} \rfloor$.)

1. Suppose first that $p = \ell$. A Sylow $p$-subgroup of $G$ is conjugate to the subgroup $S$ of upper triangular matrices with ones on the diagonal. It has order $p^{\frac{n(n-1)}{2}}$ and $N_G(S)$ is the subgroup of all upper triangular matrices with determinant 1, of order $(q-1)^{n-1} p^{\frac{n(n-1)}{2}}$. A thorough description of the $p$-subgroup structure can be found in [125]. See also Proposition 6.1 below.
2. Suppose now that $p = 2 \neq \ell$. If $n = 2$, then $S$ is generalised quaternion of order equal to the 2-part of $(q^2 - 1)$. If $n \geq 3$, a sensible way to describe the structure of $S$ is to realise it as the maximal subgroup of a Sylow 2-subgroup $\hat{S}$ of $\mathrm{GL}_n(q)$

formed by the matrices with determinant 1. The structure of $\hat{S}$ is described in [47], and can be summarised as follows. If $n = 1$, then $\hat{S}$ is cyclic of order the 2-part of $q - 1$. If $n = 2$, and $q \equiv 3 \pmod 4$, then $\hat{S}$ is semi-dihedral of order $2^{a+2}$, whilst if $q \equiv 1 \pmod 4$, then $\hat{S}$ is a wreath product $C_{2^a} \wr C_2$, where $2^a$ is the highest power of 2 dividing $q + 1$ and $q - 1$ respectively. For $n \geq 3$, write $n = \sum_j 2^j$ in its 2-adic expansion (finitely many nonzero $j$). Accordingly, $\hat{S}$ is a direct product $\prod_j S_j$ where $S_j \in \mathrm{Syl}_2(\mathrm{GL}_{2^j}(q))$ is isomorphic to a Sylow 2-subgroup of $\mathrm{GL}_2(q) \wr \mathfrak{S}_{2^{j-1}}$ for $j \geq 2$.

3. Suppose that $p$ is odd and that $q \equiv 1 \pmod p$. Let $p^a$ be the highest power of $p$ dividing $q - 1$. We can take $S = E \rtimes Q$, where $E$ is homocyclic of exponent $p^a$ and rank $n - 1$, embedded as a diagonal subgroup in $G$, and $Q$ is isomorphic to a Sylow $p$-subgroup of $\mathfrak{S}_n$ (possibly $Q$ is trivial), formed by permutation matrices. There is an exact sequence

$$1 \longrightarrow C_G(E) \longrightarrow N_G(E) \longrightarrow \mathfrak{S}_n \longrightarrow 1 , \tag{6.1}$$

with $C_G(E) = \mathbf{T}_1^F \cong (C_{q-1})^{n-1}$ the subgroup of diagonal matrices of $G$.

4. Suppose that $q \not\equiv 0, 1 \pmod p$ and that $p$ is odd. Let $p^a$ be the highest power of $p$ dividing $q^d - 1$, where $d$ is as in Theorem 6.6. Write $n = rd + c$, with $0 \leq c < d$. Then $S = E \rtimes Q$, where $E$ is homocyclic of exponent $p^a$ and rank $r$, embedded as a block diagonal subgroup $\begin{pmatrix} d \times d & & & \\ & \ddots & & \\ & & d \times d & \\ & & & I_c \end{pmatrix}$ in $G$ with $r$ blocks of size $d$ containing a cyclic factor $C_{p^a}$ of $E$, and one identity block of size $c$, and where $Q$ is isomorphic to a Sylow $p$-subgroup of $\mathfrak{S}_r$ (possibly $Q$ is trivial), formed by matrices which permute the $r$-diagonal blocks of size $d$. Again, there is an exact sequence of the form (6.1), with $C_G(E) \cong \mathbf{T}_d^F \mathrm{SL}_c(q)$, where $\mathbf{T}_d^F$ is homocyclic of exponent $q^d - 1$ and rank $r$.

We can now describe the group of endotrivial $kG$-modules for $G$ a finite group of Lie type defined over a field of characteristic $\ell$, and $k$ a field of characteristic $p$. For convenience, we only handle the main isogeny types, as listed in [98, Table 22.1]. Historically, the case $\ell = p$ was studied more than 10 years before the case $\ell \neq p$, using a different approach [34].

Let us summarise the $p$-local properties of finite groups of Lie type for $\ell = p$ that are relevant for finding $T(G)$ in this case (cf. [34, Lemma 5.1 and Propositions 7.3 and 7.4]).

**Proposition 6.1.** *Let $G$ be a finite group of Lie type defined over $\mathbb{F}_q$ with $q = p^a$, and let $S \in \mathrm{Syl}_p(G)$.*

1. *$N_G(S)/S$ is abelian.*
2. *If $G$ has twisted Lie rank 1, then $S$ is a TI subgroup of $G$.*

3. *If the twisted Lie rank of G is at least 2, then $\mathscr{E}_{\geq 2}(G)/G$ is connected, unless G is defined over $\mathbb{F}_p$ and G is of type $A_2$, $B_2$ or $G_2$, in which case the number of connected components of $\mathscr{E}_{\geq 2}(G)/G$ is as follows.*

| Type | $p = 2$ | $p = 3$ | $p = 5$ | $p \geq 7$ |
|---|---|---|---|---|
| $A_2$ (adjoint) | 2 | 3 | 3 | 3 |
| $B_2$ | 1 | 1 | 2 | 2 |
| $G_2$ | 1 | 1 | 1 | 2 |

*If G is simply connected of type $A_2$, then the number of connected components of $\mathscr{E}_{\geq 2}(G)/G$ is equal to 3 if $p \not\equiv 1 \pmod 3$ and is equal to 5 if $p \equiv 1 \pmod 3$.*

The proof of Part 3 in Proposition 6.1 is based on an analysis of the $p$-ranks of the centralisers of the elements of order $p$ in $G$. It is relatively easy to see that if either $q > p$, or if $G$ has twisted Lie rank at least 3, then each such centraliser has $p$-rank at least 3, and therefore $G$ cannot have maximal elementary abelian $p$-subgroups of rank 2. The tedious part of the proof consists in a case by case analysis of the groups $G$ of (twisted) Lie rank 2.

If the twisted Lie rank of $G$ is 1 and $G$ is defined over the smallest possible field, then $G$ is "small". In particular, for $p = 2$ and $G$ adjoint, we have $A_1(2) = \mathrm{PGL}_2(2) \cong \mathfrak{S}_3$, ${}^2A_2(2) = \mathrm{PGU}_3(2) \cong (C_3 \times C_3) \rtimes \mathrm{SL}_2(3)$, and ${}^2B_2(2) \cong C_5 \rtimes C_4$. In all the other cases, Sylow $p$-subgroups of $G$ are neither cyclic, quaternion, nor semi-dihedral (cf. also [68, Theorems 4.10.2 and 4.10.5]), implying that $TT(G) = K(G)$ in most finite groups of Lie type.

The facts necessary to calculate the torsionfree rank of $T(G)$ when the twisted Lie rank of $G$ is greater than 1 are proved in [34, Sect. 7].

Proposition 6.1 suggests that we should divide the problem of finding $T(G)$ into:

- twisted Lie rank 1, and
- twisted Lie rank greater than 1.

We start with the first case. From Proposition 6.1, we know that a Sylow $p$-subgroup $S$ is a trivial intersection subgroup of $G$, and that the quotient group $N_G(S)/S$ is abelian.

**Theorem 6.7.** ([34, Sect. 5]) *Let G be a finite group of Lie type defined over $\mathbb{F}_q$ with $q = p^a$, let $S \in \mathrm{Syl}_p(G)$ and write $N = N_G(S)$. Suppose that G has twisted Lie rank 1.*

1. *The restriction map $\mathrm{Res}_N^G : T(G) \to T(N) \cong N/S \oplus T(S)$ is an isomorphism.*
2. *If G has type $A_1(2)$, then $T(G) = \{0\}$.*
3. *If G has type $A_1(p)$, with $p > 2$, then $T(G) = TT(G)$, with $T(S) = \langle \Omega_S \rangle \cong \mathbb{Z}/2$.*
4. *If G has type ${}^2A_2(p)$, then*

$$T(G) \cong \begin{cases} \mathbb{Z}/3 \oplus T(S) & \text{if } p = 2, \\ \mathbb{Z}/(p^2 - 1) \oplus \mathbb{Z} & \text{if } p \equiv -1 \pmod{3}, \ p \text{ odd and } G \cong \mathrm{PGU}_3(p), \\ \mathbb{Z}/(p^2 - 1) \oplus \mathbb{Z}^3 & \text{if } p \equiv -1 \pmod{3}, \ p \text{ odd and } G \cong \mathrm{SU}_3(p), \\ \mathbb{Z}/(p^2 - 1) \oplus \mathbb{Z} & \text{if } p \not\equiv -1 \pmod{3}, \ p \text{ odd.} \end{cases}$$

*Moreover, if $p = 2$, then $S = Q_8$ and $T(S) \cong \mathbb{Z}/2 \oplus \mathbb{Z}/4$ since $\mathbb{F}_4$ contains a primitive cube root of 1. If $p$ is odd, then $S$ is extraspecial of order $p^3$ and exponent $p$, and so $T(S) \cong \mathbb{Z}^{p+1}$, as described in Sect. 3.3.*

5. *If $G$ has type $^2B_2(2)$, then $T(G) \cong T(S) = \langle \Omega_S \rangle \cong \mathbb{Z}/2$.*
6. *In all the other cases, $T(S) = \langle \Omega_S \rangle \cong \mathbb{Z}$.*

For the groups of twisted Lie rank greater than 1, the classification of endotrivial modules boils down to a few specific considerations when $k = \mathbb{F}_p$ and the twisted Lie rank is 2, as listed in Proposition 6.1. For such a group $G$, the torsion subgroup $TT(G)$ of $T(G)$ is generated by the stable isomorphism classes of the trivial Sylow restriction $kG$-modules. Indeed, $TT(S) = \{0\}$ because $S$ is not cyclic, quaternion, or semi-dihedral. The approach taken in [34, Theorem 6.2] is the traditional method based on the Green correspondence and uses "well-known" facts from Lie theory.

**Theorem 6.8.** *Let $G$ be a finite group of Lie type defined over $\mathbb{F}_q$ with $q = p^a$, let $S \in \mathrm{Syl}_p(G)$ and write $N = N_G(S)$. Suppose that $G$ has twisted Lie rank greater than 1. Then $TT(G) = K(G) \cong G/G'S$ is generated by the stable isomorphism classes of the 1-dimensional $kG$-modules. In other words, the $kG$-Green correspondent $M$ of a 1-dimensional $kN$-module is endotrivial if and only if $\dim(M) = 1$.*

*Proof.* First suppose that $G$ is perfect. Let $M$ be the $kG$-Green correspondent of a 1-dimensional $kN$-module $\chi$, and suppose that $M$ is endotrivial. We want to show that $\chi = k$. By assumption, $M\downarrow_N^G \cong \chi \oplus (\mathrm{proj})$. Let $n$ be the twisted Lie rank of $G$, and choose a parabolic subgroup $P_i$ corresponding to a simple root of the twisted root system, in the sense of [68, Sects. 2.2 and 2.6]. By transitivity of the restriction map, $M\downarrow_N^G \cong (M\downarrow_{P_i}^G)\downarrow_N^{P_i}$, with $M\downarrow_{P_i}^G = V \oplus (\mathrm{proj})$ for some indecomposable endotrivial $kP_i$-module $V$. By Green's correspondence, we also have $V | \chi\uparrow_N^{P_i}$, and using Mackey's formula and the Bruhat decomposition (cf. [68, Sect. 2.3]), we deduce that

$$V\downarrow_{P_i}^G \mid (\chi\uparrow_N^{P_i})\downarrow_N^{P_i} \cong \chi \oplus \,^w\chi\uparrow_{^wN \cap N}^N,$$

where $\{1, w\}$ is a set of double coset representatives of $N \backslash P_i / N$, which has cardinality 2 because $P_i$ corresponds to a simple root.

Thus $V\downarrow_N^{P_i} \mid \chi \oplus \,^w\chi\uparrow_{^wN \cap N}^N$ for some $w \in P_i$ that does not belong to $N$. Since $^wN \cap N$ is a proper subgroup of $N$ of order divisible by $p$ and which does not contain $S$, we must have $V\downarrow_N^{P_i} \cong \chi$.

So we can write $\chi = (\chi_1, \ldots, \chi_n) \in \mathrm{Hom}(N/S, k^\times)$, where $\chi_i = \chi\downarrow_{L_i'}^N$, with $L_i'$ the derived subgroup of the Levi subgroup corresponding to $P_i$, and where $n$ is the twisted Lie rank of $G$. The above argument says that we can extend $\chi$ to a $kP_i$-module $V$ for each $i$. This property is known to hold if and only if $\chi_i$ is the trivial

$kL'_i$-module for all $1 \leq i \leq n$. Therefore, $V$ is the $kP_i$-Green correspondent of the trivial $kN$-module, i.e. $V = k$, and a fortiori, $M = k$.

Now let us assume that $G$ is not perfect. So $G$ is adjoint and $G'$ is perfect. From [98, Theorem 24.17], we gather that if $p = 2$, then the only possibilities for $G$ simply connected and $G \neq G'$ are groups whose Sylow 2-subgroups are selfnormalising, and with no maximal elementary abelian 2-subgroup of rank 2. If $p$ is odd, then $G'$ is a normal subgroup of $G$ of index prime to $p$, and $G'$ is perfect. So, the above argument shows that the only 1-dimensional $kG'$-module is the trivial one, i.e. $K(G') = \{0\}$. Now, we apply Theorem 5.1, and conclude that $K(G) \cong G/G'S$, as required. $\square$

Combining Proposition 6.1 and Theorem 6.8, and using the algebra software Magma [15] for the excluded cases in these results, we obtain the following.

**Theorem 6.9.** *Let $G$ be a finite group of Lie type defined over $\mathbb{F}_q$ with $q = p^a$, let $S \in \mathrm{Syl}_p(G)$ and write $N = N_G(S)$. Suppose that $G$ has twisted Lie rank greater than 1, and that $G$ is defined over $\mathbb{F}_q$ with $q = p^a$. Then $T(G) = \langle \Omega_G \rangle \oplus K(G)$, with $K(G) \cong G/G'S$ and $TF(G) = \langle \Omega_G \rangle$ infinite cyclic, unless one of the following holds.*

1. *If $G$ is of type $A_2(p)$, then $TF(G) \cong \mathbb{Z}^2$ if $p = 2$, and $TF(G) \cong \mathbb{Z}^3$ if $p \geq 3$.*
2. *If $G$ is of type $B_2(p)$ and $p \geq 5$, then $TF(G) \cong \mathbb{Z}^2$.*
3. *If $G$ is of type $G_2(p)$ and $p \geq 7$, then $TF(G) \cong \mathbb{Z}^2$.*

*Moreover, a complete set of generators for $TF(G)$ can be specified using Theorem 4.3.*

This completes the description of the group of endotrivial modules for a finite group of Lie type in defining characteristic.

We now turn to the question of determining $T(G)$ for the finite groups $G$ of Lie type in nondefining characteristic, that is, when $\ell \neq p$. We continue with the settings of the previous section.

Our strategy to determine $K(G)$ consists in working with the algebraic group $\mathbf{G}$ (where $G = \mathbf{G}^F$, for a Steinberg endomorphism $F$ of $\mathbf{G}$), and using the known properties of complete root data (6.2). The reason is that the normaliser of a Sylow $p$-subgroup of $G$ is contained in the fixed points of the normaliser of a maximal $\Phi_d$-torus, but for a handful of exceptions (as one would expect with these groups!). Consequently, much of the $p$-local structure of $G$ can be deduced from the "Lie toolbag" that comes alongside complete root data, such as complex reflection groups and $d$-split Levi subgroups. The ubiquitous exceptions that we need to handle separately come from two main results: the cases when the normaliser of a Sylow $p$-subgroup of $G$ is not contained in the $F$-fixed points of the normaliser of a maximal $\Phi_d$-torus (cf. [97, Theorems 5.14 and 5.19] and [98, Theorem 25.20]), and Tits' theorem [98, Theorem 24.17], which we recall below.

**Notation 6.10.** *For distinct prime numbers $p$ and $\ell$, and for an $\ell$-power $q$, define $d = d_p(q)$ to be the multiplicative order of $q$ modulo $p$ if $p$ is odd, and if $p = 2$, set $d = 1$ if $q \equiv 1 \pmod 4$, and $d = 2$ if $q \equiv 3 \pmod 4$.*

**Theorem 6.11.** *Let* **G** *be a simple linear algebraic group defined over* $\mathbb{F}_q$ *with Steinberg endomorphism* $F$ *and let* $G = \mathbf{G}^F$ *and* $S \in \mathrm{Syl}_p(G)$.

1. *Suppose that* $p$ *divides* $|G|$ *and* $p \neq 2, \ell$. *Then there exists a maximal* $\Phi_d$-*torus* $\mathbf{T}_d$ *of* **G** *such that* $N_G(S) \leq N_G(\mathbf{T}_d)$ *unless* $p = 3$ *and* $G$ *is one of*

    - $\mathrm{SL}_3(q)$ *with* $q \equiv 4, 7 \pmod 9$, *or*
    - $\mathrm{SU}_3(q)$ *with* $q \equiv 2, 5 \pmod 9$, *or*
    - $G_2(q)$ *with* $q \equiv 2, 4, 5, 7 \pmod 9$.

2. *Suppose that* $p = 2$. *Then* $N_G(S) \leq N_G(\mathbf{T}_d)$ *unless*

    - $G = \mathrm{SL}_2(q)$ *with* $q \equiv 3, 5 \pmod 8$, *or*
    - $G = \mathrm{Sp}_{2n}(q)$ *with* $q \equiv 3, 5 \pmod 8$.

3. *Parts 1 and 2 can be extended to semisimple groups as follows: if* **G** *is semisimple and* $F$ *not very twisted, then* $N_G(S) \leq N_G(\mathbf{T}_d)$ *if* $p \geq 5$.
4. *Suppose that* **G** *is simply connected simple. Then* $G$ *is perfect unless* $G$ *is one of* $\mathrm{SL}_2(2)$, $\mathrm{SL}_2(3)$, $\mathrm{SU}_3(2)$, $\mathrm{Sp}_4(2)$, $G_2(2)$, $^2B_2(2)$, $^2G_2(3)$, *or* $^2F_4(2)$. *Consequently, if* $G$ *is perfect, then* $G/Z(G)$ *is simple.*

The $p$-local structure of $G$ is well known and can be found in particular in [68, Sect. 4.10], and also [47, 66, 98]. One of the main properties is that if $p$ is odd, then a Sylow $p$-subgroup of $G$ has a unique elementary abelian $p$-subgroup $E$ of maximal rank, and every elementary abelian $p$-subgroup is conjugate to a subgroup of $E$, except for a few exceptions (of course!).

In addition to the small groups identified after Proposition 6.1, we record the following isomorphisms: $\mathrm{SL}_2(3) \cong Q_8 \rtimes C_3$, $\mathrm{SU}_3(2) \cong 3^{1+2}_+ \rtimes Q_8$, $\mathrm{SL}_4(2) \cong A_8$, $\mathrm{Sp}_4(2) \cong \mathfrak{S}_6$, $G_2(2)$ has derived subgroup $\mathrm{PSU}_3(3)$ of index 2, $^2G_2(3)$ has derived subgroup $\mathrm{PSL}_2(8)$ of index 3, and $\mathrm{Sp}_{2n}(q) \cong \mathrm{SO}_{2n+1}(q)$ for $q$ even and for all $n \geq 2$.

To find $K(G)$ for $G$ a finite group of Lie type when $N_G(S) \leq N_G(\mathbf{T}_d)$, the homotopy theory method of [70] is well suited, especially [70, Theorems A and E], respectively Theorems 5.5 and 5.8. Loosely, these results allow us to use $p$-local information on the finite group $G$ in order to obtain information on the trivial Sylow restriction $kG$-modules. The computations of $K(G)$ using Theorem 5.8 resemble those carried out applying Theorem 5.4, in that we aim to show that $N_G(S)$, or $N_G(E)$ (if $E$ is characteristic in $S$, as mentioned above), is generated by elements which belong to the commutator subgroup of the centraliser of an element of order $p$. In addition, we can restrict even further the structure of the group $K(G)$ of trivial Sylow restriction $kG$-modules, deducing a nontrivial fact from the results of Steinberg on torsion primes (cf. [116]), and Tits recalled above in Theorem 6.11 Part 4.

**Proposition 6.2.** *Let* **G** *be a simple linear algebraic group defined over* $\mathbb{F}_q$ *with Steinberg endomorphism* $F$ *and let* $G = \mathbf{G}^F$ *and* $S \in \mathrm{Syl}_p(G)$. *Write* **D** *for the corresponding complete root datum, and* $\pi_1(\mathbf{D})_F$ *for the* $F$-*coinvariants of the fundamental group of* **D** *(cf. [98, Lemma 22.1]). Suppose that* **D** *has* $\Phi_d$-*rank at least 2. Then, there is an isomorphism* $(\pi_1(\mathbf{D})_F)_{p'} \to \mathrm{H}_1(\mathcal{O}_p^*(G); k^\times)_{\ell'}$. *Moreover,*

$H_1(\mathbf{D}; k^\times)_\ell = 0$ *unless $G$ is simply connected of type listed in Theorem 6.11 Part 4. Consequently, $K(G)$ can only have $\ell$-torsion.*

In many cases in the present situation, the subgroup $E$ introduced above is the unique maximal elementary abelian subgroup of $S$ of maximal rank. That is, in terms of $\Phi_d$-tori and the above notation, $E$ is the characteristic subgroup of $\mathbf{T}_d^F$ formed by the elements of order $p$ (of the Sylow $p$-subgroup) of $\mathbf{T}_d^F$, while the centraliser of an element of order $p$ is the finite group of $F$-fixed points of the $d$-split Levi subgroup corresponding to a rank one $\Phi_d$-subtorus of $\mathbf{T}_d$. The structure of the $d$-split Levi subgroups can be gathered from the root system.

*Example 6.2.* Consider the root system of type $E_7$, and say we want to find the 4-split Levi subgroups corresponding to the maximal $\Phi_4$-torus $\mathbf{T}_4$, and a rank 1 $\Phi_4$-subtorus $U$ of $\mathbf{T}_4$. Since $\Phi_4$ has multiplicity 2 in the polynomial order of $\mathbf{G}$ of type $E_7$, $\mathbf{T}_4$ has rank 2 as a torus. As integral lattices, $\mathbf{T}_4$ and $U$ have rank 4 and 2 respectively, since $\deg(\Phi_4) = 2$. Now, in the canonical basis $\{\varepsilon_j \mid 1 \le j \le 8\}$ of $\mathbb{R}^8$ the simple roots of $E_7$ are

$$\alpha_1 = \frac{1}{2}\Big(\varepsilon_1 + \varepsilon_8 - \sum_{j=2}^{7} \varepsilon_j\Big), \quad \alpha_2 = \varepsilon_1 + \varepsilon_2, \text{ and } \alpha_j = \varepsilon_{j-1} - \varepsilon_{j-2},$$

for $3 \le j \le 7$. So we can realise $\mathbf{T}_4$ and $U$ as complete root data for the root lattice on the first four and two coordinates respectively. Then the roots centralising $\mathbf{T}_4$ are $\pm\varepsilon_5 \pm \varepsilon_6$, and $\pm(\varepsilon_7 - \varepsilon_8)$, which give the 4-split Levi subgroup of type "$\Phi_4^2.A_1^3$", where the factors do not necessarily split. In fact, the above roots give $C_{\mathbf{G}}(\mathbf{T}_4) \cong \mathbf{T}_4 \times D_2 \times A_1$, where $\pm\varepsilon_5 \pm \varepsilon_6$ give $D_2$, and $\pm(\varepsilon_7 - \varepsilon_8)$ give $A_1$. The roots centralising $U$ are $\pm\varepsilon_i \pm \varepsilon_j$ with $3 \le i < j \le 6$, and $\pm(\varepsilon_7 - \varepsilon_8)$, and so, $C_{\mathbf{G}}(U) \cong U \times D_4 \times A_1$.

The Weyl group $W_4$ of $\mathbf{T}_4$ is the complex reflection group labeled $G_8$, isomorphic to a central extension $4.\mathfrak{S}_4$, and the Weyl group of $U$ is cyclic of order 4. The two cyclic factors of $\mathbf{T}_4$ are conjugate by an element of $W_4$, which means that $U$ represents the unique conjugacy class of rank one $\Phi_4$-subtori of $\mathbf{T}_4$. Now, the factor $A_1$ is common to both centralisers $C_{\mathbf{G}}(\mathbf{T}_4)$ and $C_{\mathbf{G}}(U)$, and therefore centralises any $\Phi_4$-subtorus of $\mathbf{T}_4$,

To calculate $H^1(\mathcal{O}_5^*(G); k^\times)$ using Theorem 5.8, we calculate the kernel

$$K = \ker\big(\theta : H_1(N_G([U < S]); k^\times) \to H_1(N_G(U); k^\times)\big)$$

of the map induced by the inclusion $N_G([U < S]) \le N_G(U)$, where $U = \mathbf{U}^F$ and $S = \mathbf{T}_4^F$. Recall that $H_1(X; k^\times)$ is the $p'$-part of the abelianisation $X/X'$ of a finite group $X$. We have $H_1(\mathcal{O}_5^*(G); k^\times) = H_1(N_G(S); k^\times)/K$, by identifying $K$ with a subgroup of $H_1(N_G(S); k^\times)$ (cf. also Theorem 5.9). We note that only the primes $\ell = 2$ and $\ell = 3$ can produce some nonzero outcome by Tits theorem (cf. Theorem 6.11 Part 4), and moreover, only some $\ell$-torsion in these homology groups can produce some nontrivial trivial Sylow restriction $kG$-module by Proposition 6.2.

If $\ell = 2$, we calculate $H_1(N_G(S); k^\times) = H_1(N_G([U < S]); k^\times) \cong (\mathbb{Z}/2)^3$, and $H_1(N_G(U); k^\times) \cong \mathbb{Z}/2$, where the image of $\theta$ contains the two copies of $\mathbb{Z}/2$ coming from the Klein four group obtained as the abelianisation of the group of type $D_2$, i.e. $\mathrm{Spin}_4^+(2)$ in the simply connected group and $(\mathrm{PCO}_4^0)^+(2)$ in the adjoint one. The last factor $\mathbb{Z}/2$ comes from the abelianisation of the factor of type $A_1(2)$ (isomorphic to $\mathfrak{S}_3$). We conclude that $H_1(\mathscr{O}_5^*(E_7(2)); k^\times) \cong \mathbb{Z}/2$.

If $\ell = 3$, we calculate $H_1(N_G(S); k^\times) = H_1(N_G([U < S]); k^\times) = H_1(N_G(U); k^\times)$ is the $5'$-part of the abelianisation of the finite group of type $A_1(3)$. So we obtain that $K(G)$ is cyclic of order 3 if $\mathbf{G}$ is simply connected, since $\mathrm{SL}_2(3) \cong Q_8 \rtimes C_3$ has abelianisation of order 3, whilst $K(G)$ is trivial if $\mathbf{G}$ is adjoint, because $\mathrm{PGL}_2(3) \cong \mathfrak{S}_4$ has abelianisation of order 2.

Proceeding case by case, and reducing the number of possible groups for which we may have $K(G) \not\cong G/G'S$, as illustrated in Example 6.2, we obtain a list of the finite groups of Lie type $G$ of simply connected and adjoint types, with $p$-rank at least 2 with $K(G) \not\cong G/G'S$ (cf. [32]).

**Theorem 6.12.** *Let $G$ be a finite group of Lie type defined over $\mathbb{F}_q$ with $q$ not divisible by $p$, the characteristic of $k$. Then the group of trivial Sylow restriction $kG$-modules $K(G)$ is isomorphic to $G/G'S$, the group of 1-dimensional $kG$-modules, unless one of the following holds.*

| $G$ | $p$ | $S$ | $K(G)/(G/G'S)$ | notes |
|---|---|---|---|---|
| $\mathrm{SU}_3(2)$ | 3 | $3_+^{1+2}$ | $\mathbb{Z}/2 \oplus \mathbb{Z}/2$ | $\mathrm{SU}_3(2) \cong 3_+^{1+2} \rtimes Q_8$ |
| $\mathrm{SL}_4(2)$ | 3 | $C_3 \times C_3$ | $\mathbb{Z}/2$ | $\mathrm{SL}_4(2) \cong A_8$ |
| $\mathrm{Sp}_4(2)$ | 3 | $C_3 \times C_3$ | $\mathbb{Z}/2$ | $\mathrm{Sp}_4(2) \cong \mathfrak{S}_6$ |
| $G_2(2)$ | 3 | $3_+^{1+2}$ | $\mathbb{Z}/2$ | $S$ is TI |
| $^2G_2(3)$ | 2 | $(C_2)^3$ | $\mathbb{Z}/3$ | $S$ is TI |
| $F_4(2)$ | 5 | $C_5 \times C_5$ | $\mathbb{Z}/2$ | |
| $^2F_4(2)$ | 5 | $C_5 \times C_5$ | $\mathbb{Z}/4$ | $S$ is TI |
| $\mathrm{Sp}_8(2)$ | 5 | $C_5 \times C_5$ | $\mathbb{Z}/2$ | |
| $E_7(2)$ | 5 | $C_5 \times C_5$ | $\mathbb{Z}/2$ | Example 6.2 |
| $E_7(3)$ simply connected | 5 | $C_5 \times C_5$ | $\mathbb{Z}/3$ | Example 6.2 |

For completeness, let us also mention that:

- Since $G_2(2)' \cong \mathrm{PSU}_3(3)$ has index 2 in $G_2(2)$, if $p = 3$, then $K(G) \cong K(N) \cong \mathbb{Z}/8$ because $S$ is TI (cf. Theorem 6.9).
- Since $^2G_2(3)$ has derived subgroup $\mathrm{PSL}_2(8)$ of index 3, if $p = 2$, then $K(G) \cong G/G'S \cong \mathbb{Z}/3$, and $\mathrm{PSL}_2(8)$ has a TI Sylow 2-subgroup with index 7 in its normaliser (cf. Theorem 6.9).
- The Tits group $^2F_4(2)'$ has index 2 in $F_4(2)$, and it is dealt with in Sect. 6.3.

To find $TF(G)$, we need to calculate the number of connected components of $\mathscr{E}_{\geq 2}(G)/G$, which is often a connected poset (and so $TF(G) = \langle \Omega_G \rangle \cong \mathbb{Z}$). But there are cases for which the torsionfree rank is greater, and these depend on the isogeny type of $G$. To see when such situations occur, we rely on a result of Steinberg on *torsion primes*, that is, the prime divisors of torsion elements in the fundamental group of **G** (see in particular [116, Theorems 2.27 and 2.28]). In addition, we know from Theorem 3.4 that we only need to consider the finite groups of $p$-rank at most $p$ if $p$ is odd and at most 4 if $p = 2$.

**Theorem 6.13.** *Let $G$ be be a finite group of Lie type defined over $\mathbb{F}_q$ of simply connected or adjoint type, and suppose that $q$ is not divisible by $p$, the characteristic of $k$. The torsionfree part of the group of endotrivial $kG$-modules is infinite cyclic unless one of the following hold.*

1. *$p = 2$ and $G$ has type $A_1$, in which case, if $G = SL_2(q)$, then $TF(G) = \{0\}$, and if $G = PGL_2(q)$, then $TF(G) \cong \mathbb{Z}^2$. (See Sects. 3.10 and 3.8.)*
2. *Suppose that $p$ is odd.*

   a. *$G$ has cyclic Sylow $p$-subgroups, in which case $TF(G) = \{0\}$,*
   b. *$G = {}^3D_4(q)$ with $p = 3$, in which case $TF(G) \cong \mathbb{Z}^2$,*
   c. *$G \cong PGL_p(q)$ with $q \equiv 1 \pmod{p}$, in which case $TF(G) \cong \mathbb{Z}^3$,*
   d. *$G \cong PGU_p(q)$ with $q \equiv -1 \pmod{p}$, in which case $TF(G) \cong \mathbb{Z}^3$.*

## 6.3 Endotrivial Modules for Sporadic Groups and Their Covering Groups

There are twenty-six sporadic simple groups, as detailed in [68, Sects. 5.3-5] and in [49], for instance. Their Schur multipliers are described in [68, Sect. 6.1], and the upshot is that they are cyclic groups of order dividing 12. In fact, only the Schur multipliers of $M_{22}$, Suz and $Fi_{22}$ have order greater than 3.

Let $G$ be a covering group of a sporadic simple group. That is, $G = G'$ is quasi-simple with $G/Z(G)$ simple sporadic. Let $S \in Syl_p(G)$ and $N = N_G(S)$.

In Sects. 6.3.1 and 6.3.2, we give the structure of $T(G)$ for each prime divisor of $|G|$ in the cases when $S$ is noncyclic and cyclic, respectively. The proofs can be found in [90], where the results are obtained by a case by case inspection of each group and prime. We use a blend of the techniques described in Chap. 5 to find $K(G)$, and also computer calculations with GAP [62] and Magma [15].

The torsionfree rank of $T(G)$ is the "easy" part of the exercise, because the $p$-local structure of the simple sporadic groups and their covering groups is well known, and in case of a gap, then the algebra software Magma [15] comes in handy. Let us also point out that the results for the Monster sporadic group $F_1$, missing in [90], have been obtained by Grodal in [70, Theorem 6.1].

### 6.3.1   $T(G)$ When rank$(S) \geq 2$

In order to find the group of endotrivial modules $T(G)$ when $G$ has $p$-rank at least 2, we split the problem into a relatively straightforward computation of the $\mathbb{Z}$-rank of $TF(G)$ (i.e. the number of connected components of $\mathscr{E}_{\geq 2}(G)/G$), and the analysis of $K(G)$, mainly using the character theory method from Sect. 5.1 and the algebra software GAP. We also rely on the known facts on the group structure of the sporadic simple groups as can be found in the Atlas [49]. We use the notation of the Atlas.

Let us give as example the cases of the Mathieu groups $M_{11}$ and $M_{12}$, both at the prime $p = 3$.

A Sylow 3-subgroup $S$ of $G = M_{11}$ is elementary abelian of order 9, and $S$ is a TI subgroup of $G$, with normaliser $N \cong S \rtimes SD_{16}$, where $SD_{16}$ is a semi-dihedral 2-group of order 16. So $TF(G) = \langle \Omega_G \rangle \cong \mathbb{Z}$ and $N/N'S \cong (\mathbb{Z}/2)^2$ is a Klein four group. By Theorem 5.1, we conclude that $TT(G) = K(G) \cong (\mathbb{Z}/2)^2$. Explicitly, $K(G)$ is generated by the $kG$-Green correspondents of the 1-dimensional $kN$-modules, which we can calculate with Magma [15]. These have dimensions 1, 55, 55 and 10.

A Sylow 3-subgroup $S$ of $G = M_{12}$ is extraspecial of order 27 and exponent 3, with normaliser $N \cong S \rtimes (C_2 \times C_2)$. The four maximal elementary abelian subgroups of $S$ fuse into two $G$-conjugacy classes, giving that $TF(G) \cong \mathbb{Z}^2$. Now, $N$ is contained in a subgroup $H = (C_3 \times C_3) \rtimes \mathrm{GL}_2(3)$. Since $O_3(H) \neq 1$, we have $K(H) \cong H/H'S \cong \mathbb{Z}/2$. The restriction $\mathrm{Res}_H^G : K(G) \to K(H)$ being injective, it suffices to consider the nontrivial 1-dimensional $kH$-module and its $kG$-Green correspondent. Using the method of Sect. 5.1, if $\chi$ denotes the nontrivial 1-dimensional $kH$-module, then we calculate with GAP the part $e_0 \cdot \mathrm{Ind}_H^G(\chi)$ of the induced character that lies in the principal block. From the output, we gather that the $kG$-Green correspondent $\Gamma^G(\chi)$ of $\chi$ affords the character $\chi_9 + \chi_{13}$ of degree $175 \not\equiv 1 \pmod{|S|}$. Therefore $\Gamma^G(\chi)$ cannot be endotrivial and we conclude that $K(G) = \{0\}$.

Now, for the covering groups, note that if $m.G$ is a covering group of $G$ and $p$ does not divide $m$, then $\mathrm{Inf}_G^{m.G} : T(G) \to T(m.G)$ is injective.

Proceeding in this fashion group by group and prime by prime, we obtain an almost complete classification of endotrivial modules for the finite sporadic simple groups and their covering groups at all the primes $p$ for which these groups have $p$-rank at least 2.

We summarise these results in the table below, where an empty entry means that rank$(S) \leq 1$. An isomorphism "$\cong T(G)$" means "isomorphic via inflation". A question mark indicates that only a partial result for the structure of $TT(G)$ has been obtained, and in Sect. 6.3.1.1, we give $N/N'S$ as a maximal bound for $TT(G)$. A question mark bounded by a group, e.g. $(? \leq \mathbb{Z}/2)$ for the group B in characteristic 5, indicates that we have found a sharper bound for $TT(G)$ than the one given by $N/N'S$.

| $G$ | $p = 2$ | $p = 3$ | $p = 5$ |
|-----|---------|---------|---------|
| $J_2$ | $\mathbb{Z}$ | $\mathbb{Z} \oplus \mathbb{Z}/2$ | $\mathbb{Z} \oplus \mathbb{Z}/2$ |
| $2.J_2$ | $\mathbb{Z}$ | $\cong T(J_2)$ | $\cong T(J_2)$ |
| $^2F_4(2)'$ | $\mathbb{Z}$ | $\mathbb{Z}^2$ | $\mathbb{Z} \oplus \mathbb{Z}/6$ |
| HS | $\mathbb{Z}$ | $\mathbb{Z} \oplus (\mathbb{Z}/2)^2$ | $\mathbb{Z}^2 \oplus \mathbb{Z}/4$ |
| 2.HS | $\mathbb{Z}$ | $\cong T(\text{HS})$ | $\cong T(\text{HS})$ |
| McL | $\mathbb{Z}$ | $\mathbb{Z}$ | $\mathbb{Z} \oplus \mathbb{Z}/8$ |
| 3.McL | $\mathbb{Z}$ | $\mathbb{Z}$ | $\mathbb{Z} \oplus \mathbb{Z}/24$ |
| Ru | $\mathbb{Z}$ | $\mathbb{Z} \oplus \mathbb{Z}/2$ | $\mathbb{Z}^2 \oplus ?$ |
| 2.Ru | $\mathbb{Z}$ | $\mathbb{Z} \oplus \mathbb{Z}/4$ | $\cong T(\text{Ru})$ |
| Suz | $\mathbb{Z}$ | $\mathbb{Z}$ | $\mathbb{Z} \oplus \mathbb{Z}/2$ |
| 2.Suz | $\mathbb{Z}$ | $\cong T(\text{Suz})$ | $\cong T(\text{Suz})$ |
| 3.Suz | $\mathbb{Z}$ | $\mathbb{Z}$ | $\cong T(\text{Suz})$ |
| 6.Suz | $\mathbb{Z}$ | $\mathbb{Z}$ | $\cong T(\text{Suz})$ |
| $Co_3$ | $\mathbb{Z}$ | $\mathbb{Z}$ | $\mathbb{Z} \oplus ?$ |
| $Co_2$ | $\mathbb{Z}$ | $\mathbb{Z}$ | $\mathbb{Z} \oplus ?$ |
| $Fi_{22}$ | $\mathbb{Z}$ | $\mathbb{Z}$ | $\mathbb{Z} \oplus \mathbb{Z}/2$ |
| $2.Fi_{22}$ | $\mathbb{Z}$ | $\cong T(Fi_{22})$ | $\mathbb{Z} \oplus (\mathbb{Z}/2)^2$ |
| $3.Fi_{22}$ | $\mathbb{Z}$ | $\mathbb{Z}$ | $\mathbb{Z} \oplus \mathbb{Z}/6$ |
| $6.Fi_{22}$ | $\mathbb{Z}$ | $\mathbb{Z}$ | $\mathbb{Z} \oplus \mathbb{Z}/6 \oplus \mathbb{Z}/2$ |
| $Fi_{23}$ | $\mathbb{Z}$ | $\mathbb{Z} \oplus ?$ | $\mathbb{Z} \oplus \mathbb{Z}/2$ |

| $G$ | $p = 2$ | $p = 3$ |
|-----|---------|---------|
| $M_{11}$ | $\mathbb{Z} \oplus \mathbb{Z}/2$ | $\mathbb{Z} \oplus (\mathbb{Z}/2)^2$ |
| $M_{12}$ | $\mathbb{Z}$ | $\mathbb{Z}^3$ |
| $2.M_{12}$ | $\mathbb{Z}$ | $\cong T(M_{12})$ |
| $M_{22}$ | $\mathbb{Z}$ | $\mathbb{Z} \oplus (\mathbb{Z}/2)^2$ |
| $2.M_{22}$ | $\mathbb{Z}$ | $\mathbb{Z} \oplus \mathbb{Z}/2 \oplus \mathbb{Z}/4$ |
| $3.M_{22}$ | $\mathbb{Z}$ | $\mathbb{Z}$ |
| $4.M_{22}$ | $\mathbb{Z}$ | $\cong T(2.M_{22})$ |
| $6.M_{22}$ | $\mathbb{Z}$ | $\mathbb{Z}$ |
| $12.M_{22}$ | $\mathbb{Z}$ | $\mathbb{Z}$ |
| $M_{23}$ | $\mathbb{Z}$ | $\mathbb{Z} \oplus \mathbb{Z}/2$ |
| $M_{24}$ | $\mathbb{Z}$ | $\mathbb{Z}^2$ |
| $J_1$ | $\mathbb{Z}$ | |
| $J_3$ | $\mathbb{Z}$ | $\mathbb{Z} \oplus (? \leq \mathbb{Z}/2)$ |
| $3.J_3$ | $\mathbb{Z} \oplus \mathbb{Z}/3$ | $\mathbb{Z}$ |

| $G$ | $p = 2$ | $p = 3$ | $p = 5$ | $p = 7$ |
|-----|---------|---------|---------|---------|
| O'N | $\mathbb{Z}$ | $\mathbb{Z} \oplus (? \leq \mathbb{Z}/2)$ | | $\mathbb{Z}^3$ |
| 3.O'N | $\mathbb{Z}$ | $\mathbb{Z}$ | | $\mathbb{Z}^3$ |
| He | $\mathbb{Z}$ | $\mathbb{Z}^2$ | $\mathbb{Z} \oplus \mathbb{Z}/3$ | $\mathbb{Z}^3 \oplus ?$ |
| $F_5 = HN$ | $\mathbb{Z}$ | $\mathbb{Z}$ | $\mathbb{Z}$ | |
| Ly | $\mathbb{Z}$ | $\mathbb{Z}$ | $\mathbb{Z}$ | |
| Th | $\mathbb{Z}$ | $\mathbb{Z} \oplus ?$ | $\mathbb{Z} \oplus ?$ | $\mathbb{Z} \oplus ?$ |
| $Co_1$ | $\mathbb{Z}$ | $\mathbb{Z}$ | $\mathbb{Z}^2 \oplus (? \leq \mathbb{Z}/4)$ | $\mathbb{Z} \oplus ?$ |
| $2.Co_1$ | $\mathbb{Z}$ | $\cong T(Co_1)$ | $\cong T(Co_1)$ | $\cong T(Co_1)$ |
| $Fi_{24}'$ | $\mathbb{Z}$ | $\mathbb{Z} \oplus ?$ | $\mathbb{Z} \oplus ?$ | $\mathbb{Z}^3 \oplus ?$ |
| $3.Fi_{24}'$ | $\mathbb{Z}$ | $\mathbb{Z}$ | $\cong T(Fi_{24}')$ | $\cong T(Fi_{24}')$ |
| $F_2 = B$ | $\mathbb{Z}$ | $\mathbb{Z} \oplus ?$ | $\mathbb{Z} \oplus (? \leq \mathbb{Z}/2)$ | $\mathbb{Z} \oplus ?$ |
| $2.F_2 = 2.B$ | $\mathbb{Z}$ | $\cong T(B)$ | $\cong T(B)$ | $\cong T(B)$ |

| $G$ | $p=2$ | $p=3$ | $p=5$ | $p=7$ | $p=11$ | $p=13$ |
|---|---|---|---|---|---|---|
| $J_4$ | $\mathbb{Z}$ | $\mathbb{Z}$ | | | $\mathbb{Z} \oplus \mathbb{Z}/10$ | |
| $F_1 = M$ | $\mathbb{Z}$ | $\mathbb{Z}$ | $\mathbb{Z}$ | $\mathbb{Z}$ | $\mathbb{Z}$ | $\mathbb{Z}^2$ |

#### 6.3.1.1 The Structure of $N/N'S$ for the Missing Cases Above

We have listed all the triples $(G, p, N/N'S)$ for which there is a question mark in the above table and the best bound for $K(G)$ is $N/N'S$.

| $G$ | $p$ | $N/N'S$ | $G$ | $p$ | $N/N'S$ | $G$ | $p$ | $N/N'S$ |
|---|---|---|---|---|---|---|---|---|
| O'N | 3 | $\mathbb{Z}/2$ | Ru | 5 | $\mathbb{Z}/2 \oplus \mathbb{Z}/4$ | He | 7 | $\mathbb{Z}/6$ |
| Th | 3 | $(\mathbb{Z}/2)^2$ | $Co_3$ | 5 | $\mathbb{Z}/2 \oplus \mathbb{Z}/4$ | Th | 7 | $\mathbb{Z}/6$ |
| $Fi_{23}$ | 3 | $(\mathbb{Z}/2)^3$ | $Co_2$ | 5 | $\mathbb{Z}/4$ | $Co_1$ | 7 | $\mathbb{Z}/3 \oplus \mathbb{Z}/3$ |
| $Fi'_{24}$ | 3 | $(\mathbb{Z}/2)^3$ | Th | 5 | $\mathbb{Z}/4$ | $Fi'_{24}$ | 7 | $\mathbb{Z}/2 \oplus \mathbb{Z}/6$ |
| B | 3 | $(\mathbb{Z}/2)^3$ | $Fi'_{24}$ | 5 | $\mathbb{Z}/4$ | B | 7 | $\mathbb{Z}/2 \oplus \mathbb{Z}/6$ |

### 6.3.2 $T(G)$ for Covering Groups of Sporadic Groups with Cyclic Sylow $p$-Subgroups

By [49], we know that if $S$ is cyclic, then $|S| = p$ and $p$ is odd. In particular, $S$ is TI, so that $\mathrm{Res}_H^G : T(G) \to T(N)$ is an isomorphism, with $T(N)/\langle \Omega_N \rangle \cong N/N'S \cong \mathbb{Z}/e$, where $e = |N : C_G(S)|$ is the inertial index of the principal block of $kG$. By [104], $T(G) \cong N/N'S \oplus \mathbb{Z}/2$ if and only if there exists a 1-dimensional $kN$-module $\chi$ such that $\chi \cong \Omega_N^2(k)$, which happens in particular when $e$ is odd. Otherwise, $T(G) = \langle \Omega_G \rangle \cong \mathbb{Z}/2e$. The results are obtained by a routine computation with the algebra softwares GAP [62] and Magma [15], and analysing the structure of $N$ using the standard resources such as the Atlas of finite simple groups [49], and Gorenstein, Lyons and Solomon's Volume 3 [68]. We summarise the results in the following tables.

#### 6.3.2.1 The Mathieu Groups

| $G$ | $p$ | $N/N'S$ | $e$ | $T(G)$ | $G$ | $p$ | $N/N'S$ | $e$ | $T(G)$ |
|---|---|---|---|---|---|---|---|---|---|
| $M_{11}$ | 5 | $\mathbb{Z}/4$ | 4 | $\mathbb{Z}/8$ | $M_{11}$ | 11 | $\mathbb{Z}/5$ | 5 | $\mathbb{Z}/10$ |
| $M_{12}$ | 5 | $\mathbb{Z}/2 \oplus \mathbb{Z}/4$ | 4 | $\mathbb{Z}/2 \oplus \mathbb{Z}/8$ | $M_{12}$ | 11 | $\mathbb{Z}/5$ | 5 | $\mathbb{Z}/10$ |
| $2.M_{12}$ | 5 | $\mathbb{Z}/2 \oplus \mathbb{Z}/4$ | 4 | $\cong T(M_{12})$ | $2.M_{12}$ | 11 | $\mathbb{Z}/10$ | 5 | $\mathbb{Z}/2 \oplus \mathbb{Z}/10$ |

| $G$ | $p$ | $N/N'S$ | $e$ | $T(G)$ | $G$ | $p$ | $N/N'S$ | $e$ | $T(G)$ |
|---|---|---|---|---|---|---|---|---|---|
| $M_{22}$ | 5 | $\mathbb{Z}/4$ | 4 | $\mathbb{Z}/8$ | $2.M_{22}$ | 5 | $\mathbb{Z}/2 \oplus \mathbb{Z}/4$ | 4 | $\mathbb{Z}/2 \oplus \mathbb{Z}/8$ |
| $3.M_{22}$ | 5 | $\mathbb{Z}/12$ | 4 | $\mathbb{Z}/24$ | $4.M_{22}$ | 5 | $\mathbb{Z}/2 \oplus \mathbb{Z}/8$ | 4 | $\mathbb{Z}/4 \oplus \mathbb{Z}/8$ |
| $6.M_{22}$ | 5 | $\mathbb{Z}/2 \oplus \mathbb{Z}/12$ | 4 | $\mathbb{Z}/2 \oplus \mathbb{Z}/24$ | $12.M_{22}$ | 5 | $\mathbb{Z}/2 \oplus \mathbb{Z}/24$ | 4 | $\mathbb{Z}/4 \oplus \mathbb{Z}/24$ |
| $M_{22}$ | 7 | $\mathbb{Z}/3$ | 3 | $\mathbb{Z}/6$ | $2.M_{22}$ | 7 | $\mathbb{Z}/6$ | 3 | $\mathbb{Z}/2 \oplus \mathbb{Z}/6$ |
| $3.M_{22}$ | 7 | $\mathbb{Z}/3 \oplus \mathbb{Z}/3$ | 3 | $\mathbb{Z}/3 \oplus \mathbb{Z}/6$ | $4.M_{22}$ | 7 | $\mathbb{Z}/12$ | 3 | $\mathbb{Z}/2 \oplus \mathbb{Z}/12$ |
| $6.M_{22}$ | 7 | $\mathbb{Z}/3 \oplus \mathbb{Z}/6$ | 3 | $\mathbb{Z}/6 \oplus \mathbb{Z}/6$ | $12.M_{22}$ | 7 | $\mathbb{Z}/3 \oplus \mathbb{Z}/12$ | 3 | $\mathbb{Z}/6 \oplus \mathbb{Z}/12$ |
| $M_{22}$ | 11 | $\mathbb{Z}/5$ | 5 | $\mathbb{Z}/10$ | $2.M_{22}$ | 11 | $\mathbb{Z}/10$ | 5 | $\mathbb{Z}/2 \oplus \mathbb{Z}/10$ |
| $3.M_{22}$ | 11 | $\mathbb{Z}/15$ | 5 | $\mathbb{Z}/30$ | $4.M_{22}$ | 11 | $\mathbb{Z}/20$ | 5 | $\mathbb{Z}/2 \oplus \mathbb{Z}/20$ |
| $6.M_{22}$ | 11 | $\mathbb{Z}/30$ | 5 | $\mathbb{Z}/2 \oplus \mathbb{Z}/30$ | $12.M_{22}$ | 11 | $\mathbb{Z}/60$ | 5 | $\mathbb{Z}/2 \oplus \mathbb{Z}/60$ |
| $M_{23}$ | 5 | $\mathbb{Z}/4$ | 4 | $\mathbb{Z}/8$ | $M_{23}$ | 7 | $\mathbb{Z}/6$ | 3 | $\mathbb{Z}/2 \oplus \mathbb{Z}/6$ |
| $M_{23}$ | 11 | $\mathbb{Z}/5$ | 5 | $\mathbb{Z}/10$ | $M_{23}$ | 23 | $\mathbb{Z}/11$ | 11 | $\mathbb{Z}/22$ |
| $M_{24}$ | 5 | $\mathbb{Z}/4$ | 4 | $\mathbb{Z}/8$ | $M_{24}$ | 7 | $\mathbb{Z}/6$ | 3 | $\mathbb{Z}/2 \oplus \mathbb{Z}/6$ |
| $M_{24}$ | 11 | $\mathbb{Z}/10$ | 10 | $\mathbb{Z}/20$ | $M_{24}$ | 23 | $\mathbb{Z}/11$ | 11 | $\mathbb{Z}/22$ |

#### 6.3.2.2 The Janko Groups

| $G$ | $p$ | $N/N'S$ | $e$ | $T(G)$ | $G$ | $p$ | $N/N'S$ | $e$ | $T(G)$ |
|---|---|---|---|---|---|---|---|---|---|
| $J_1$ | 3 | $\mathbb{Z}/2 \oplus \mathbb{Z}/2$ | 2 | $\mathbb{Z}/2 \oplus \mathbb{Z}/4$ | $J_1$ | 5 | $\mathbb{Z}/2 \oplus \mathbb{Z}/2$ | 2 | $\mathbb{Z}/2 \oplus \mathbb{Z}/4$ |
| $J_1$ | 7 | $\mathbb{Z}/6$ | 6 | $\mathbb{Z}/12$ | $J_1$ | 11 | $\mathbb{Z}/10$ | 10 | $\mathbb{Z}/20$ |
| $J_1$ | 19 | $\mathbb{Z}/6$ | 6 | $\mathbb{Z}/12$ | | | | | |
| $J_2$ | 7 | $\mathbb{Z}/6$ | 6 | $\mathbb{Z}/12$ | $2.J_2$ | 7 | $\mathbb{Z}/12$ | 6 | $\mathbb{Z}/2 \oplus \mathbb{Z}/12$ |
| $J_3$ | 5 | $\mathbb{Z}/2 \oplus \mathbb{Z}/2$ | 2 | $\mathbb{Z}/2 \oplus \mathbb{Z}/4$ | $3.J_3$ | 5 | $\mathbb{Z}/2 \oplus \mathbb{Z}/6$ | 2 | $\mathbb{Z}/2 \oplus \mathbb{Z}/12$ |
| $J_3$ | 17 | $\mathbb{Z}/8$ | 8 | $\mathbb{Z}/16$ | $3.J_3$ | 17 | $\mathbb{Z}/24$ | 8 | $\mathbb{Z}/48$ |
| $J_3$ | 19 | $\mathbb{Z}/9$ | 9 | $\mathbb{Z}/18$ | $3.J_3$ | 19 | $\mathbb{Z}/3 \oplus \mathbb{Z}/9$ | 9 | $\mathbb{Z}/3 \oplus \mathbb{Z}/18$ |
| $J_4$ | 5 | $\mathbb{Z}/4$ | 4 | $\mathbb{Z}/8$ | $J_4$ | 7 | $\mathbb{Z}/6$ | 3 | $\mathbb{Z}/2 \oplus \mathbb{Z}/6$ |
| $J_4$ | 23 | $\mathbb{Z}/22$ | 22 | $\mathbb{Z}/44$ | $J_4$ | 29 | $\mathbb{Z}/28$ | 28 | $\mathbb{Z}/56$ |
| $J_4$ | 31 | $\mathbb{Z}/10$ | 10 | $\mathbb{Z}/20$ | $J_4$ | 37 | $\mathbb{Z}/12$ | 12 | $\mathbb{Z}/24$ |
| $J_4$ | 43 | $\mathbb{Z}/14$ | 14 | $\mathbb{Z}/28$ | | | | | |

### 6.3.2.3 The Conway Groups

| $G$ | $p$ | $N/N'S$ | $e$ | $T(G)$ | $G$ | $p$ | $N/N'S$ | $e$ | $T(G)$ |
|---|---|---|---|---|---|---|---|---|---|
| $Co_3$ | 7 | $\mathbb{Z}/2\oplus\mathbb{Z}/6$ | 6 | $\mathbb{Z}/2\oplus\mathbb{Z}/12$ | $Co_3$ | 11 | $\mathbb{Z}/10$ | 5 | $\mathbb{Z}/2\oplus\mathbb{Z}/10$ |
| $Co_3$ | 23 | $\mathbb{Z}/11$ | 11 | $\mathbb{Z}/22$ | $Co_2$ | 7 | $\mathbb{Z}/2\oplus\mathbb{Z}/6$ | 6 | $\mathbb{Z}/2\oplus\mathbb{Z}/12$ |
| $Co_2$ | 11 | $\mathbb{Z}/10$ | 10 | $\mathbb{Z}/20$ | $Co_2$ | 23 | $\mathbb{Z}/11$ | 11 | $\mathbb{Z}/22$ |
| $Co_1$ | 11 | $\mathbb{Z}/2\oplus\mathbb{Z}/10$ | 10 | $\mathbb{Z}/2\oplus\mathbb{Z}/20$ | $2.Co_1$ | 11 | $\mathbb{Z}/2\oplus\mathbb{Z}/10$ | 10 | $\cong T(Co_1)$ |
| $Co_1$ | 13 | $\mathbb{Z}/12$ | 12 | $\mathbb{Z}/24$ | $2.Co_1$ | 13 | $\mathbb{Z}/12$ | 12 | $\cong T(Co_1)$ |
| $Co_1$ | 23 | $\mathbb{Z}/11$ | 11 | $\mathbb{Z}/22$ | $2.Co_1$ | 23 | $\mathbb{Z}/22$ | 11 | $\mathbb{Z}/2\oplus\mathbb{Z}/22$ |

### 6.3.2.4 The Suzuki Groups

| $G$ | $p$ | $N/N'S$ | $e$ | $T(G)$ | $G$ | $p$ | $N/N'S$ | $e$ | $T(G)$ |
|---|---|---|---|---|---|---|---|---|---|
| Suz | 7 | $\mathbb{Z}/6$ | 6 | $\mathbb{Z}/12$ | 2.Suz | 7 | $\mathbb{Z}/6$ | 6 | $\cong T(\text{Suz})$ |
| 3.Suz | 7 | $\mathbb{Z}/3\oplus\mathbb{Z}/6$ | 6 | $\mathbb{Z}/3\oplus\mathbb{Z}/12$ | 6.Suz | 7 | $\mathbb{Z}/3\oplus\mathbb{Z}/6$ | 6 | $\cong T(3.\text{Suz})$ |
| Suz | 11 | $\mathbb{Z}/10$ | 10 | $\mathbb{Z}/20$ | 2.Suz | 11 | $\mathbb{Z}/20$ | 10 | $\mathbb{Z}/2\oplus\mathbb{Z}/20$ |
| 3.Suz | 11 | $\mathbb{Z}/30$ | 10 | $\mathbb{Z}/60$ | 6.Suz | 11 | $\mathbb{Z}/60$ | 10 | $\mathbb{Z}/2\oplus\mathbb{Z}/60$ |
| Suz | 13 | $\mathbb{Z}/6$ | 6 | $\mathbb{Z}/12$ | 2.Suz | 13 | $\mathbb{Z}/12$ | 6 | $\mathbb{Z}/2\oplus\mathbb{Z}/12$ |
| 3.Suz | 13 | $\mathbb{Z}/3\oplus\mathbb{Z}/6$ | 6 | $\mathbb{Z}/3\oplus\mathbb{Z}/12$ | 6.Suz | 13 | $\mathbb{Z}/3\oplus\mathbb{Z}/12$ | 6 | $\mathbb{Z}/6\oplus\mathbb{Z}/12$ |

### 6.3.2.5 The Fischer Groups

| $G$ | $p$ | $N/N'S$ | $e$ | $T(G)$ | $G$ | $p$ | $N/N'S$ | $e$ | $T(G)$ |
|---|---|---|---|---|---|---|---|---|---|
| $Fi_{22}$ | 7 | $\mathbb{Z}/2\oplus\mathbb{Z}/6$ | 6 | $\mathbb{Z}/2\oplus\mathbb{Z}/12$ | $2.Fi_{22}$ | 7 | $(\mathbb{Z}/2)^2\oplus\mathbb{Z}/6$ | 6 | $(\mathbb{Z}/2)^2\oplus\mathbb{Z}/12$ |
| $3.Fi_{22}$ | 7 | $(\mathbb{Z}/6)^2$ | 6 | $\mathbb{Z}/6\oplus\mathbb{Z}/12$ | $6.Fi_{22}$ | 7 | $\mathbb{Z}/2\oplus(\mathbb{Z}/6)^2$ | 6 | $\mathbb{Z}/2\oplus\mathbb{Z}/6\oplus\mathbb{Z}/12$ |
| $Fi_{22}$ | 11 | $\mathbb{Z}/10$ | 5 | $\mathbb{Z}/2\oplus\mathbb{Z}/10$ | $2.Fi_{22}$ | 11 | $\mathbb{Z}/2\oplus\mathbb{Z}/10$ | 5 | $(\mathbb{Z}/2)^2\oplus\mathbb{Z}/10$ |
| $3.Fi_{22}$ | 11 | $\mathbb{Z}/30$ | 5 | $\mathbb{Z}/2\oplus\mathbb{Z}/30$ | $6.Fi_{22}$ | 11 | $\mathbb{Z}/2\oplus\mathbb{Z}/30$ | 5 | $(\mathbb{Z}/2)^2\oplus\mathbb{Z}/30$ |
| $Fi_{22}$ | 13 | $\mathbb{Z}/6$ | 6 | $\mathbb{Z}/12$ | $2.Fi_{22}$ | 13 | $\mathbb{Z}/2\oplus\mathbb{Z}/6$ | 6 | $\mathbb{Z}/2\oplus\mathbb{Z}/12$ |
| $3.Fi_{22}$ | 13 | $\mathbb{Z}/3\oplus\mathbb{Z}/6$ | 6 | $\mathbb{Z}/3\oplus\mathbb{Z}/12$ | $6.Fi_{22}$ | 13 | $\mathbb{Z}/6\oplus\mathbb{Z}/6$ | 6 | $\mathbb{Z}/6\oplus\mathbb{Z}/12$ |
| $Fi_{23}$ | 7 | $\mathbb{Z}/2\oplus\mathbb{Z}/6$ | 6 | $\mathbb{Z}/2\oplus\mathbb{Z}/12$ | $Fi_{23}$ | 11 | $\mathbb{Z}/2\oplus\mathbb{Z}/10$ | 10 | $\mathbb{Z}/2\oplus\mathbb{Z}/20$ |
| $Fi_{23}$ | 13 | $\mathbb{Z}/2\oplus\mathbb{Z}/6$ | 6 | $\mathbb{Z}/2\oplus\mathbb{Z}/12$ | $Fi_{23}$ | 17 | $\mathbb{Z}/16$ | 16 | $\mathbb{Z}/32$ |
| $Fi_{23}$ | 23 | $\mathbb{Z}/11$ | 11 | $\mathbb{Z}/22$ |  |  |  |  |  |
| $Fi'_{24}$ | 11 | $\mathbb{Z}/10$ | 10 | $\mathbb{Z}/20$ | $3.Fi'_{24}$ | 11 | $\mathbb{Z}/30$ | 10 | $\mathbb{Z}/60$ |
| $Fi'_{24}$ | 13 | $\mathbb{Z}/2\oplus\mathbb{Z}/12$ | 12 | $\mathbb{Z}/2\oplus\mathbb{Z}/24$ | $3.Fi'_{24}$ | 13 | $\mathbb{Z}/2\oplus\mathbb{Z}/12$ | 12 | $\cong T(Fi'_{24})$ |
| $Fi'_{24}$ | 17 | $\mathbb{Z}/16$ | 16 | $\mathbb{Z}/32$ | $3.Fi'_{24}$ | 17 | $\mathbb{Z}/48$ | 16 | $\mathbb{Z}/96$ |
| $Fi'_{24}$ | 23 | $\mathbb{Z}/11$ | 11 | $\mathbb{Z}/22$ | $3.Fi'_{24}$ | 23 | $\mathbb{Z}/33$ | 11 | $\mathbb{Z}/66$ |
| $Fi'_{24}$ | 29 | $\mathbb{Z}/14$ | 14 | $\mathbb{Z}/28$ | $3.Fi'_{24}$ | 29 | $\mathbb{Z}/42$ | 14 | $\mathbb{Z}/84$ |

### 6.3.2.6 The Remaining Groups

| $G$ | $p$ | $N/N'S$ | $e$ | $T(G)$ | $G$ | $p$ | $N/N'S$ | $e$ | $T(G)$ |
|---|---|---|---|---|---|---|---|---|---|
| $^2F_4(2)'$ | 13 | $\mathbb{Z}/6$ | 6 | $\mathbb{Z}/12$ | He | 17 | $\mathbb{Z}/8$ | 8 | $\mathbb{Z}/16$ |
| Ru | 7 | $\mathbb{Z}/6$ | 6 | $\mathbb{Z}/12$ | 2.Ru | 7 | $\mathbb{Z}/6$ | 6 | $\cong T(\text{Ru})$ |
| Ru | 13 | $\mathbb{Z}/12$ | 12 | $\mathbb{Z}/24$ | 2.Ru | 13 | $\mathbb{Z}/12$ | 12 | $\cong T(\text{Ru})$ |
| Ru | 29 | $\mathbb{Z}/14$ | 14 | $\mathbb{Z}/28$ | 2.Ru | 29 | $\mathbb{Z}/28$ | 14 | $\mathbb{Z}/2 \oplus \mathbb{Z}/28$ |
| ON | 5 | $\mathbb{Z}/2 \oplus \mathbb{Z}/4$ | 4 | $\mathbb{Z}/2 \oplus \mathbb{Z}/8$ | 3.ON | 5 | $\mathbb{Z}/2 \oplus \mathbb{Z}/4$ | 4 | $\cong T(\text{ON})$ |
| ON | 11 | $\mathbb{Z}/10$ | 10 | $\mathbb{Z}/20$ | 3.ON | 11 | $\mathbb{Z}/30$ | 10 | $\mathbb{Z}/60$ |
| ON | 19 | $\mathbb{Z}/6$ | 6 | $\mathbb{Z}/12$ | 3.ON | 19 | $\mathbb{Z}/3 \oplus \mathbb{Z}/6$ | 6 | $\mathbb{Z}/3 \oplus \mathbb{Z}/12$ |
| ON | 31 | $\mathbb{Z}/15$ | 15 | $\mathbb{Z}/30$ | 3.ON | 31 | $\mathbb{Z}/3 \oplus \mathbb{Z}/15$ | 15 | $\mathbb{Z}/3 \oplus \mathbb{Z}/30$ |
| HS | 7 | $\mathbb{Z}/6$ | 6 | $\mathbb{Z}/12$ | 2.HS | 7 | $\mathbb{Z}/12$ | 6 | $\mathbb{Z}/2 \oplus \mathbb{Z}/12$ |
| HS | 11 | $\mathbb{Z}/5$ | 5 | $\mathbb{Z}/10$ | 2.HS | 11 | $\mathbb{Z}/10$ | 5 | $\mathbb{Z}/2 \oplus \mathbb{Z}/10$ |

| $G$ | $p$ | $N/N'S$ | $e$ | $T(G)$ | $G$ | $p$ | $N/N'S$ | $e$ | $T(G)$ |
|---|---|---|---|---|---|---|---|---|---|
| HN | 7 | $\mathbb{Z}/6$ | 6 | $\mathbb{Z}/12$ | HN | 11 | $\mathbb{Z}/2 \oplus \mathbb{Z}/10$ | 10 | $\mathbb{Z}/2 \oplus \mathbb{Z}/20$ |
| HN | 19 | $\mathbb{Z}/9$ | 9 | $\mathbb{Z}/18$ | | | | | |
| Ly | 7 | $\mathbb{Z}/6$ | 6 | $\mathbb{Z}/12$ | Ly | 11 | $\mathbb{Z}/10$ | 5 | $\mathbb{Z}/2 \oplus \mathbb{Z}/10$ |
| Ly | 31 | $\mathbb{Z}/6$ | 6 | $\mathbb{Z}/12$ | Ly | 37 | $\mathbb{Z}/18$ | 18 | $\mathbb{Z}/36$ |
| Ly | 67 | $\mathbb{Z}/22$ | 22 | $\mathbb{Z}/44$ | | | | | |
| McL | 7 | $\mathbb{Z}/6$ | 3 | $\mathbb{Z}/2 \oplus \mathbb{Z}/6$ | 3.McL | 7 | $\mathbb{Z}/3 \oplus \mathbb{Z}/6$ | 3 | $(\mathbb{Z}/6)^2$ |
| McL | 11 | $\mathbb{Z}/5$ | 5 | $\mathbb{Z}/10$ | 3.McL | 11 | $\mathbb{Z}/15$ | 5 | $\mathbb{Z}/30$ |
| Th | 13 | $\mathbb{Z}/12$ | 12 | $\mathbb{Z}/24$ | Th | 19 | $\mathbb{Z}/18$ | 18 | $\mathbb{Z}/36$ |
| Th | 31 | $\mathbb{Z}/15$ | 15 | $\mathbb{Z}/30$ | | | | | |
| B | 11 | $\mathbb{Z}/2 \oplus \mathbb{Z}/10$ | 10 | $\mathbb{Z}/2 \oplus \mathbb{Z}/20$ | 2.B | 11 | $\mathbb{Z}/2 \oplus \mathbb{Z}/10$ | 10 | $\cong T(\text{B})$ |
| B | 13 | $\mathbb{Z}/2 \oplus \mathbb{Z}/12$ | 12 | $\mathbb{Z}/2 \oplus \mathbb{Z}/24$ | 2.B | 13 | $\mathbb{Z}/2 \oplus \mathbb{Z}/12$ | 12 | $\cong T(\text{B})$ |
| B | 17 | $\mathbb{Z}/2 \oplus \mathbb{Z}/16$ | 16 | $\mathbb{Z}/2 \oplus \mathbb{Z}/32$ | 2.B | 17 | $\mathbb{Z}/2 \oplus \mathbb{Z}/16$ | 16 | $\cong T(\text{B})$ |
| B | 19 | $\mathbb{Z}/2 \oplus \mathbb{Z}/18$ | 18 | $\mathbb{Z}/2 \oplus \mathbb{Z}/36$ | 2.B | 19 | $\mathbb{Z}/2 \oplus \mathbb{Z}/18$ | 18 | $\cong T(\text{B})$ |
| B | 23 | $\mathbb{Z}/22$ | 11 | $\mathbb{Z}/2 \oplus \mathbb{Z}/22$ | 2.B | 23 | $\mathbb{Z}/2 \oplus \mathbb{Z}/22$ | 11 | $(\mathbb{Z}/2)^2 \oplus \mathbb{Z}/22$ |
| B | 31 | $\mathbb{Z}/15$ | 15 | $\mathbb{Z}/30$ | 2.B | 31 | $\mathbb{Z}/30$ | 15 | $\mathbb{Z}/2 \oplus \mathbb{Z}/30$ |
| B | 47 | $\mathbb{Z}/23$ | 23 | $\mathbb{Z}/46$ | 2.B | 47 | $\mathbb{Z}/46$ | 23 | $\mathbb{Z}/2 \oplus \mathbb{Z}/46$ |
| M | 17 | $\mathbb{Z}/16$ | 16 | $\mathbb{Z}/32$ | M | 19 | $\mathbb{Z}/18$ | 18 | $\mathbb{Z}/36$ |
| M | 23 | $\mathbb{Z}/22$ | 11 | $\mathbb{Z}/2 \oplus \mathbb{Z}/22$ | M | 29 | $\mathbb{Z}/28$ | 28 | $\mathbb{Z}/56$ |
| M | 31 | $\mathbb{Z}/30$ | 15 | $\mathbb{Z}/2 \oplus \mathbb{Z}/30$ | M | 41 | $\mathbb{Z}/40$ | 40 | $\mathbb{Z}/80$ |
| M | 47 | $\mathbb{Z}/46$ | 23 | $\mathbb{Z}/2 \oplus \mathbb{Z}/46$ | M | 59 | $\mathbb{Z}/29$ | 29 | $\mathbb{Z}/58$ |
| M | 71 | $\mathbb{Z}/35$ | 35 | $\mathbb{Z}/70$ | | | | | |

## 6.4   A Last Observation… Conjecture?

Section 6.3 concludes our review of endotrivial modules and their classification. Considering the number of groups discussed, for each relevant prime, we are perplexed by the lack of nontrivial trivial Sylow restriction $kG$-modules when the finite group $G$ has no strongly $p$-embedded subgroup and "large" $p$-rank. This absence of instances leads us to the following open question.

*The last question.* Let $G$ be a finite group, let $k$ be a field of characteristic $p$ and let $S \in \mathrm{Syl}_p(G)$. Suppose that $S$ has rank at least 3 and that $G$ has no proper strongly $p$-embedded subgroup. Do we necessarily have $K(G) = G/G'S$?

# References

1. J.L. Alperin, *Local Representation Theory*. Cambridge Studies in Advanced Mathematics, vol. 11 (Cambridge University Press, Cambridge, 1986)
2. J.L. Alperin, Invertible modules for groups. Notices Am. Math. Soc. **24** (1977)
3. J.L. Alperin, Construction of endo-permutation modules. J. Group Theory **4**, 3–10 (2001)
4. J.L. Alperin, Lifting endo-trivial modules. J. Group Theory **4**, 1–2 (2001)
5. M. Auslander, I. Reiten, Representation theory of Artin algebras IV. Invariants given by almost split sequences. Comm. Algebra **5**, 443–518 (1977)
6. M. Auslander, I. Reiten, *Representation Theory of Artin Algebras*. Cambridge Studies in Advanced Mathematics, vol. 36 (Cambridge University Press, Cambridge, 1995)
7. M. Auslander, J. Carlson, Almost-split sequences and group rings. J. Algebra **103**(1), 122–140 (1986)
8. P. Balmer, Modular representations of finite groups with trivial restriction to Sylow subgroups. J. Eur. Math. Soc. **15**, 2061–2079 (2013)
9. T. Barthel, J. Grodal, J. Hunt, Torsionfree endotrivial modules via homotopy theory, in preparation
10. D.J. Benson, *Modular Representation Theory: New Trends and Methods* (Springer, New York, 1984)
11. D.J. Benson, *Representations and Cohomology, I and II*. Cambridge Studies in Advanced Mathematics 30 and 31 (Cambridge University Press, Cambridge, 1991)
12. D.J. Benson, J.F. Carlson, Nilpotent elements in the Green ring. J. Algebra **104**(2), 329–350 (1986)
13. C. Bessenrodt, Endotrivial modules and the Auslander–Reiten quiver, in *Representation Theory of Finite Groups and Finite-Dimensional Algebras*, Progr. Math. 95, (Birkhäuser, Basel, 1991), pp. 317–326
14. N. Blackburn, Groups of prime-power order having an abelian centralizer of type $(r, 1)$. Mh. Math. **99**, 1–18 (1985)
15. W. Bosma, J. Cannon, C. Playoust, The Magma algebra system. I. The user language. J. Symbolic Comput. **24**, 235–265 (1997)
16. S. Bouc, *Biset Functors for Finite Groups*, vol. 1990. Lecture Notes in Mathematics (Springer, Heidelberg, 2010)
17. S. Bouc, Tensor induction of relative syzygies. J. Reine Angew. Math. **523**, 113–171 (2000)
18. S. Bouc, The Dade group of a $p$-group. Inv. Math. **164**, 189–231 (2006)
19. S. Bouc, Gluing endo-permutation modules. J. Group Theory **12**(5), 651–678 (2009)
20. S. Bouc, N. Mazza, The Dade group of (almost) extraspecial $p$-groups. J. Pure Appl. Algebra **192**, 21–51 (2004)
21. S. Bouc, J. Thévenaz, The group of endo-permutation modules. Inv. Math. **139**, 275–349 (2000)

© The Author(s), under exclusive license to Springer Nature Switzerland AG 2019
N. Mazza, *Endotrivial Modules*, SpringerBriefs in Mathematics,
https://doi.org/10.1007/978-3-030-18156-7

22. G.E. Bredon, *Topology and Geometry*. Graduate Texts in Mathematics, vol.139 (Springer, New York, 1993)
23. M. Broue, On Scott modules and p-permutation modules: an approach through the Brauer morphism. Proc. Am. Math. Soc. **93**, 401–408 (1985)
24. M. Broué, G. Malle, J. Michel, Représentations Unipotentes Génériques et Blocs des Groupes Réductifs Finis. Astérisque **212** (1993)
25. K.S. Brown, *Cohomology of Groups*. Graduate Texts in Mathematics, vol. 87 (Springer, New York, 1994)
26. J.F. Carlson, *Modules and Group Algebras* (Birkhäuser, Basel, 1996)
27. J.F. Carlson, Constructing endotrivial modules. J. Pure Appl. Algebra **1–2**, 83–110 (2006)
28. J.F. Carlson, Maximal elementary abelian subgroups of rank 2. J. Group Theory **10**, 5–14 (2007)
29. J.F. Carlson, *Cohomology and Representation Theory. Group Representation Theory* (EPFL Press, Lausanne, 2007), pp. 3–45
30. J.F. Carlson, Endotrivial modules, in *Proceedings of Symposia in Pure Mathematics Recent Developments in Lie Algebras, Groups and Representation Theory*, vol. 86 (American Mathematical Society, Providence, 2012), pp. 99–111
31. J.F. Carlson, Toward a classification of endotrivial modules, in *Finite Simple Groups: Thirty Years of the Atlas and Beyond*. Contemporary Mathematics, vol. 694, (American Mathematical Society, Providence, 2017), pp. 139–150
32. J.F. Carlson, J. Grodal, N. Mazza, D. Nakano, *Endotrivial Modules for Finite Reductive Groups*, preprint
33. J.F. Carlson, D. Hemmer, N. Mazza, Endotrivial modules for the symmetric and alternating groups. Proc. Edinb. Math. Soc. **53**(01), 83–95 (2010)
34. J.F. Carlson, N. Mazza, D. Nakano, Endotrivial modules for finite groups of Lie type. J. Reine Angew. Math. **595**, 93–120 (2006)
35. J.F. Carlson, N. Mazza, D. Nakano, Endotrivial modules for the symmetric and alternating groups. Proc. Edinb. Math. Soc. **52**(01), 45–66 (2009)
36. J.F. Carlson, N. Mazza, D. Nakano, Endotrivial modules for the general linear group in a nondefining characteristic. Math. Zeit. **278**, 901–925 (2014)
37. J.F. Carlson, N. Mazza, D. Nakano, Endotrivial modules for finite groups of Lie type *A* in a nondefining characteristic. Math. Zeit. **278**, 901–925 (2014)
38. J.F. Carlson, N. Mazza, J. Thévenaz, Endotrivial modules for *p*-solvable groups. Trans. Am. Math. Soc. **363**(9), 4979–4996 (2011)
39. J.F. Carlson, N. Mazza, J. Thévenaz, Endotrivial modules over groups with quaternion or semi-dihedral Sylow 2-subgroup. J. Eur. Math. Soc **15**, 157–177 (2013)
40. J.F. Carlson, N. Mazza, J. Thévenaz, Torsion-free endotrivial modules. J. Algebra **398**, 413–433 (2014)
41. J.F. Carlson, D. Nakano, Endotrivial modules for finite group schemes. J. Reine Angew. Math. **653**, 149–178 (2011)
42. J.F. Carlson, J. Thévenaz, Torsion endo-trivial modules. Algebras Represent. Theory **3**, 303–335 (2000)
43. J.F. Carlson, J. Thévenaz, The classification of torsion endo-trivial modules. Ann. Math. **162**(2), 823–883 (2005)
44. J.F. Carlson, J. Thévenaz, The classification of endo-trivial modules. Invent. Math. **158**(2), 389–411 (2004)
45. J.F. Carlson, J. Thévenaz, The torsion group of endotrivial modules. Algebra Number Theory **9**(3), 749–765 (2015)
46. J.F. Carlson, L. Townsend, L. Valero-Elizondo, M. Zhang, *Cohomology Rings of Finite Groups. With an Appendix: Calculations of Cohomology Rings of Groups of Order Dividing 64* (Kluwer Academic Publishers, Dordrecht, 2003)
47. R.W. Carter, P. Fong, The Sylow 2-subgroups of finite classical groups. J. Algebra **1**, 139–151 (1964)

48. R.W. Carter, *Finite Groups of Lie Type: Conjugacy Classes and Complex Characters* (Wiley, New York, 1985)

49. J.H. Conway, R.T. Curtis, S.P. Norton, R.A. Parker, R.A. Wilson, *Atlas of Finite Groups* (Clarendon Press, Oxford, 1985)

50. C. Curtis, E. Reiner, *Representation Theory of Finite Dimensional Algebras* (Wiley, New York, 1962)

51. C. Curtis, E. Reiner, *Methods of Representation Theory—With Applications to Finite Groups and Orders*, vol. I and II, (Wiley, New York, 1981 and 1987)

52. E.C. Dade, Une extension de la théorie de Hall et Highman. J. Algebra **20**, 570–609 (1972)

53. E.C. Dade, Endo-permutation modules over $p$-groups I and II. Ann. Math. **107**, 459–494 (1978) and **108**, 317–346 (1978)

54. E.C. Dade, *Extending Endo-Permutation Modules*, unpublished manuscript (1982)

55. T. Tom Dieck, *Transformation Groups*, vol. 8. De Gruyter Studies in Mathematics (Walter de Gruyter, 1987)

56. W. Dwyer, H. Henn, *Homotopy Theoretic Methods in Group Cohomology*, Advanced Courses in Mathematics CRM Barcelona (Birkhäuser, Basel, 2001)

57. K. Erdmann, *Blocks of Tame Representation Type and Related Algebras*, vol. 1428. Lectures Notes in Mathematics (Springer, Heidelberg, 1990)

58. K. Erdmann, Algebras and semidihedral defect groups I. Proc. London Math. Soc. **3**(1), 109–150 (1988)

59. K. Erdmann, Algebras and semidihedral defect groups II. Proc. Lond. Math. Soc. **3**(1), 123–165 (1990)

60. W. Feit, *The Representation Theory of Finite Groups* (North Holland Publishing Company, 1982)

61. W. Feit, Some consequences of the classification of finite simple groups. Proc. Sympos. Pure Math. **37**, 175–181 (1980)

62. The GAP Group, GAP—Groups, Algorithms, and Programming (2004)

63. G. Glauberman, Global and local properties of finite groups, in *Finite Simple Groups* (Proc. Instructional Conf., Oxford, 1969) (Academic Press, London, 1971), pp. 1–64

64. G. Glauberman, N. Mazza, $p$-groups with maximal elementary abelian subgroups of rank 2. J. Algebra **323**(6), 1729–1737 (2010)

65. D. Gorenstein, *Finite Groups*, 2nd edn. (Chelsea Publishing Co., New York, 1980)

66. D. Gorenstein, R. Lyons, The local structure of finite groups of characteristic 2 type. Mem. Am. Math. Soc. **42**(276) (1983)

67. D. Gorenstein, R. Lyons, R. Solomon, *The Classification of the Finite Simple Groups*, vol. 40, no. 2 (American Mathematical Society, 1996)

68. D. Gorenstein, R. Lyons, R. Solomon, *The Classification of the Finite Simple Groups*, vol. 40, no. 3 (American Mathematical Society, 1998)

69. J. Grodal, Higher limits via subgroup complexes. Ann. Math. **155**, 405–457 (2002)

70. J. Grodal, *Endotrivial modules for finite groups via homotopy theory*, preprint (2018)

71. A. Gullon, *On extending Scott modules*, PhD Thesis (Lancaster University, 2017)

72. M. Hall, *The Theory of Groups* (Macmillan, New York, 1959)

73. P. Hall, A contribution to the theory of groups of prime-power order. Proc. Lond. Math. Soc. **1**, 29–95 (1934)

74. P. Hall, G. Higman, On the $p$-length of $p$-soluble groups and reduction theorems for Burnside's problem. Proc. Lond. Math. Soc. **3**(6), 1–42 (1956)

75. D. Happel, *Triangulated Categories and the Representation Theory of Finite dimensional Algebras*. London Mathematical Society Lecture Note Series 119 (Cambridge University Press, Cambridge, 1988)

76. M.E. Harris, M. Linckelmann, Splendid derived equivalences for blocks of finite $p$-solvable groups. J. Lond. Math. Soc. **2**(62), 85–96 (2000)

77. A. Henke, On $p$-Kostka numbers and young modules. Eur. J. Combin. **26**(6), 923–942 (2005)

78. L. Héthelyi, Soft subgroups of $p$-groups. Ann. Univ. Sci. Budapest. Eötvös Sect. Math. **27**, 81–84 (1984)

79. L. Héthelyi, On subgroups of $p$-groups having soft subgroups. J. Lond. Math. Soc. **2**(3), 425–437 (1990)

80. P.N. Hoffman, J.F. Humphreys, *Projective Representations of the Symmetric Groups* (Oxford University Press, Oxford, 1992)

81. B. Huppert, *Endliche Gruppen I* (Springer-Verlag, Heidelberg, 1967)

82. S. Koshitani, C. Lassueur, Endo-trivial modules for finite groups with Klein-four Sylow 2-subgroups. Manuscripta Math. **148**(1–2), 265–282 (2015)

83. S. Koshitani, C. Lassueur, Endo-trivial modules for finite groups with dihedral Sylow 2-subgroup. J. Group Theory **19**(4), 635–660 (2016)

84. B. Külshammer, *Lectures on Block Theory, LMS Lecture Notes Series 161* (Cambridge University Press, Cambridge, 1991)

85. P. Landrock, *Finite Group Algebras and their Modules*. London Mathematical Society Lecture Note Series, vol. 84 (Cambridge University Press, Cambridge, 1983)

86. C. Lassueur, Relative projectivity and relative endotrivial modules. J. Algebra **337**, 285–317 (2011)

87. C. Lassueur, Relative endotrivial modules and endo-p-permutation modules in the Auslander-Reiten quiver. J. Algebra **420**, 1–14 (2014)

88. C. Lassueur, G. Malle, Simple endotrivial modules for linear, unitary and exceptional groups. Math. Z. **280**(3–4), 1047–1074 (2015)

89. C. Lassueur, G. Malle, E. Schulte, Simple endotrivial modules for Quasi-simple groups. J. Reine Angew. Math. **712**, 141–174 (2016)

90. C. Lassueur, N. Mazza, Endotrivial modules for the sporadic simple groups and their covers. J. Pure Appl. Algebra **219**, 4203–4228 (2015)

91. C. Lassueur, N. Mazza, Endotrivial modules for the Schur covers of the symmetric and alternating groups. Algebras Represent. Theory **18**(5), 1321–1335 (2015)

92. C. Lassueur, J. Thévenaz, Endotrivial modules: a reduction to $p$-central extensions. Pacific J. Math. **287**(2), 423–438 (2017)

93. M. Linckelmann, On stable equivalences with endopermutation source. J. Algebra **434**, 27–45 (2014)

94. M. Linckelmann, N. Mazza, The Dade group of a fusion system. J. Group Theory **12**, 55–74 (2009)

95. S. Mac Lane, *Categories for the Working Mathematician*. Graduate Texts in Mathematics, vol. 5 (Springer, New York, 1971)

96. A.R. MacWilliams, On 2-groups with no normal abelian subgroups of rank 3, and their occurrence as Sylow 2-subgroups of finite simple groups. Trans. Am. Math. Soc. **150**, 345–408 (1970)

97. G. Malle, Height 0 characters of finite groups of Lie type. Represent. Theory **11**, 192–220 (2007)

98. G. Malle, D. Testerman, *Linear Algebraic Groups and Finite Groups of Lie Type*. Cambridge Studies in Advanced Mathematics, vol. 133 (Cambridge University Press, Cambridge, 2011)

99. N. Mazza, *Modules d'endo-permutation* (Université de Lausanne, Switzerland, 2003). PhD Thesis

100. N. Mazza, Endo-permutation modules as sources of simple modules. J. Group Theory **6**, 477–497 (2003)

101. N. Mazza, Endotrivial modules in the normal case. J. Pure Appl. Algebra **209**, 311–323 (2007)

102. N. Mazza, Connected components of the category of elementary Abelian subgroups. J. Algebra **320**(12), 42–48 (2008)

103. N. Mazza, P. Symonds, *The Stable Category and Invertible Modules for Infinite Groups* (2019). arxiv:1902.06533

104. N. Mazza, J. Thévenaz, Endotrivial modules in the cyclic case. Archiv der Mathematik **89**(6), 497–503 (2007)

105. G. Navarro, G. Robinson, On endo-trivial modules for $p$-solvable groups. Math. Z. **270**(3–4), 983–987 (2012)

106. L. Puig, Notes sur les $P$-algèbres de Dade (Unpublished manuscript 1988)

107. L. Puig, Affirmative answer to a question of Feit. J. Algebra **131**, 513–526 (1990)
108. L. Puig, Une correspondance de modules pour les blocs à groupes de défaut abéliens. Geom. Dedic. **37**, 9–43 (1991)
109. D. Robinson, *A Course in the Theory of Groups*, 2nd edn. Graduate Texts in Mathematics, vol. 80 (Springer, New York, 1996)
110. G. Robinson, On simple endotrivial modules. Bull. Lond. Math. Soc. **43**, 712–716 (2011)
111. G. Robinson, Endotrivial irreducible lattices. J. Algebra **335**(1), 319–327 (2011)
112. J. Rotman, *An Introduction to Algebraic Topology*. Graduate Texts in Mathematics (Springer, New York, 1988)
113. J.-P. Serre, *Linear Representations of Finite Groups*. Graduate Texts in Mathematics, vol. 42 (Springer, New York, 1977)
114. S. Smith, *Subgroup Complexes*. Mathematical Surveys and Monographs, vol. 179 (American Mathematical Society, 2011)
115. T.A. Springer, R. Steinberg, Conjugacy classes, in *Seminar on Algebraic Groups Related finite Groups*, Princeton 1968/69. Lecture Notes in Mathematics (Springer, Heidelberg, 1970), pp. 167–266
116. R. Steinberg, Torsion in reductive groups. Adv. Math. **15**, 63–92 (1975)
117. A. Talian, On endotrivial modules for Lie superalgebras. J. Algebra **433**, 1–34 (2015)
118. J. Thévenaz, *G-Algebras and Modular Representation Theory* (Oxford University Press, Oxford, 1995)
119. J. Thévenaz, Relative projective covers and almost split sequences. Comm. Alg. **13**(7), 1535–1554 (1985)
120. J. Thévenaz, Endo-permutation modules, a guided tour, in *Group Representation Theory. Based on the Research Semester Group Representation Theory*, Lausanne, Switzerland, January–June, 2005, (EPFL Press, 2007), pp. 115–147
121. J.-M. Urfer, Endo-$p$-permutation modules. J. Algebra **316**(1), 206–223 (2007)
122. P. Webb, The Auslander-Reiten quiver of a finite group. Math. Z. **179**, 97–121 (1982)
123. P. Webb, Subgroup complexes, in *The Arcata Conference on Representations of Finite Groups, Proceedings of Symposia in Pure Mathematics*, vol. 47 (American Mathematical Society, 1987), pp. 349–365
124. P. Webb, An introduction to the representations and cohomology of categories, in M. Geck et al., (ed.) *Group Representation Theory. Based on the Research Semester Group Representation Theory, Lausanne, Switzerland, January–June, 2005*. Fundamental Sciences: Mathematics (EPFL Press, 2007), pp. 149–173
125. A. Weir, Sylow $p$-subgroups of the general linear group over finite fields of characteristic $p$. Proc. Am. Math. Soc. **6**, 454–464 (1955)
126. R. Wilson, *The Finite Simple Groups*. Graduate Texts in Mathematics, vol. 251 (Springer, London, 2009)

# Index

© The Author(s), under exclusive license to Springer Nature Switzerland AG 2019
N. Mazza, *Endotrivial Modules*, SpringerBriefs in Mathematics,
https://doi.org/10.1007/978-3-030-18156-7

Printed in the United States
By Bookmasters